INDIA'S GRACE CASCADES

India's Grace Cascades

WHISPERING STORIES

Ehsan Sheroy

Mohammed Altaf Hussain

Contents

Table of Content

Introduction

1. Introduce the enchanting world of India's majestic cascades.
2. Explore the cultural significance of waterfalls in Indian mythology and history.
3. Set the stage for a journey through the diverse landscapes where these waterfalls grace the land.

Chapter 1: Nature's Symphony Begins

1.1 Delve into the geological formations that give rise to India's stunning waterfalls.

1.2 Showcase iconic cascades such as Jog Falls, Athirapally Falls, and Dudhsagar Falls.

1.3 Share anecdotes and local legends associated with these natural wonders.

Chapter 2: The Elegance of Western Ghats

2.1 Focus on the Western Ghats, a hotspot for waterfalls in India.

2.2 Explore the lush greenery, biodiversity, and the symbiotic relationship between the waterfalls and the ecosystem.

2.3 Highlight lesser-known gems hidden within the Western Ghats.

Chapter 3: Northern Charms: Himalayan Cascades

3.1 Journey to the northern regions of India and explore the cascades nestled in the Himalayan foothills.

3.2 Discuss the spiritual and cultural importance of these waterfalls.

3.3 Share stories of adventure and exploration in the challenging terrains.

Chapter 4: Eastern Mystique: Waterfalls in the East

4.1 Uncover the lesser-explored waterfalls in the eastern part of India.

4.2 Discuss the unique features of these cascades and their impact on the local communities.

4.3 Explore the interplay of culture, tradition, and nature in the East.

Chapter 5: Southern Rhythms: Waterfalls in the Deccan Plateau

5.1 Investigate the waterfalls that grace the Deccan Plateau in southern India.

5.2 Discuss the geological formations and historical significance of these cascades.

5.3 Showcase the diversity of landscapes and experiences in the southern regions.

Chapter 6: Conservation and Challenges

6.1 Address the environmental challenges faced by India's waterfalls.

6.2 Explore conservation efforts and initiatives to protect these natural wonders.

6.3 Discuss the role of local communities and authorities in preserving the beauty of the cascades.

Chapter 7: The Future Flow

7.1 Reflect on the future of India's waterfalls in the face of climate change and human impact.

7.2 Discuss potential sustainable practices and responsible tourism.

7.3 Conclude with a hopeful message about the enduring grace of India's cascades and their importance for generations to

come.

Introduction

In the tremendous woven artwork of India's topographical and social scene, the fountains stand as ethereal narrators, their waters winding around stories of greatness and beauty. These regular marvels, brought into the world from the hug of different territories and land arrangements, have scratched their presence across the subcontinent, becoming vital sections in the story of India's rich legacy.

As one sets out on the excursion to unwind the mysteries held by these fountains, an orchestra of nature unfurls. Every cascade, a novel note in this fabulous structure, adds to the amicable tune that reverberates through the land. The tales murmured by these fountains are not simply the spouting hints of water meeting earth; they are stories of social importance, topographical wonders, and the fragile dance between human networks and the regular world.

The prologue to "India's Effortlessness: Fountains Murmuring Stories" welcomes perusers to drench themselves in the charm of these cascades, investigating the advantageous connection between the land and its fluid veins. Past the actual excellence, the fountains take the stand concerning the complex stories of folklore and history that have formed the Indian subcontinent.

Diving into the topographical material of India, the story leaves on an excursion that traverses the length and broadness of the country. The Western Ghats, embellished with rich vegetation and dynamic biodiversity, stand as a demonstration of the tastefulness of nature's plan. Here, cascades like Run Falls, Athirapally Falls, and Dudhsagar Falls become heroes in the unfurling adventure, their flowing waters mirroring the remarkable force of the regular world.

Wandering northwards, the account climbs into the Himalayan lower regions, where overflows cut through the rough scenes as well as hold significant profound and social importance. In the midst of the transcending tops, cascades become channels of heavenly energy, and their presence is joined with the texture of neighborhood convictions and ceremonies.

The investigation doesn't stop there; it slides toward the eastern areas, where lesser-investigated cascades uncover themselves as unexpected, yet invaluable treasures. Here, the story uncovers the persona of these fountains and their essential job in the regular routines of the networks staying in their shadows. The transaction of culture, custom, and nature becomes the dominant focal point in the lesser-voyaged corners of the eastern regions.

Yet again the excursion through India's fountains further stretches out to the Deccan Level in the south, where the scene changes, offering an alternate beat to the water's plunge. Land developments and verifiable stories entwine, illustrating the variety that characterizes the southern districts of the subcontinent.

However, in the midst of the magnificence and glory, challenges loom not too far off. The sections that continue in this account go up against the ecological issues looked by India's cascades. As the world wrestles with environmental change and human effect, these fountains demonstrate the veracity of the repercussions of uncontrolled turn of events. It is inside these moves that the story tracks down valuable chances to investigate protection endeavors, local area drives, and the sensitive equilibrium expected to guarantee the supported presence of these regular marvels.

The finishing up sections cast a look toward what's in store. What lies ahead for India's fountains in a world continually developing and confronting remarkable ecological changes? The account thinks about the job of capable the travel industry, supportable practices, and the aggregate endeavors expected to protect the tradition of these fountains for a long time into the future.

In "India's Elegance: Fountains Murmuring Stories," the presentation makes way for an odyssey through the core of the subcontinent, where cascades are not simply geographical arrangements but rather residing accounts carving stories of beauty, grandness, and the persevering through association among nature and humankind. The story coaxes perusers to listen intently, for in the murmurs of these fountains, India's accounts wake up.

1. **Introduce the enchanting world of India's majestic cascades.**

   In the huge and various scene of India, where the ensemble of nature unfurls in heap shapes, the lofty fountains stand as charming observers to the eminent excellence of the subcontinent. These cascades, brought into the world from the topographical complexities of the land, overflow down with an effortlessness that enthralls the spirit. As the fluid strings slip, they cut their ways through mountains, levels, and valleys, making an embroidery of normal ponders that describe stories of old times and significant social importance.

   The prologue to the captivating universe of India's fountains is an encouragement to set out on a tactile excursion. It coaxes the inquisitive voyager to investigate the secret stories that exist in the hurrying waters, stories that are just about ancient. These cascades, frequently celebrated in nearby legends and folklore, have become something other than topographical arrangements; they are living substances, throbbing with the heartbeat of the land.

To really see the value in the glory of India's fountains, one must initially comprehend the geographical cycles that bring about these remarkable normal marvels. From the transcending pinnacles of the Himalayas to the undulating landscape of the Western Ghats and the tremendous spread of the Deccan Level, every area adds to the different material whereupon these fountains paint their stories. It is a demonstration of the geographical wealth of the Indian subcontinent, a lavishness that shows itself as cascades that reach from delicate fountains to deafening downpours.

As the excursion through India's fountains unfurls, the story delivers the social importance woven into the texture of these scenes. In numerous fanciful stories, cascades are portrayed as holy destinations, accepted to be the residence of divinities or the appearance of heavenly energy. Travelers and enthusiasts frequently attempt difficult excursions to arrive at these hallowed fountains, looking for favors and profound restoration in the cloudy hug of the falling waters.

The Western Ghats, with its lavish vegetation and energetic biodiversity, turns into a point of convergence in the story. Here, cascades like Run Falls in Karnataka, Athirapally Falls in Kerala, and Dudhsagar Falls in Goa feature the polish of nature's plan. The flowing waters, mirroring the daylight in a heap of tones, make an exhibition that isn't just outwardly dazzling yet additionally summons a feeling of stunningness and marvel.

Moving northwards, the excursion rises to the Himalayan lower regions, where the scene is overwhelmed by transcending tops and unblemished valleys. Here, cascades are not only topographical arrangements but rather vital components in the profound and social woven artwork of the district. The Ganges, thought about a hallowed waterway in Hinduism, drops from the Himalayas with various cascades along its course, each holding a one of a kind spot in the hearts of individuals.

Wandering into the eastern locales, the account investigates less popular cascades that uncover themselves as unlikely treasures. In these more unfamiliar corners, the fountains become more than normal marvels; they are life savers for the networks that dwell in their vicinity. The interchange of culture, custom, and nature becomes the dominant focal point as the account disentangles the narratives murmured by the cascades of the East.

The investigation broadens further south, to the Deccan Level, where land arrangements and authentic stories unite. Cascades around here, like those in the Western Ghats and the Eastern Ghats, add an alternate musicality to the story. They become narrators as well as observers to the authentic occasions and social developments that have molded the southern districts of India.

However, as the story unfurls, it doesn't avoid the difficulties that go up against these fountains. The natural issues looked by India's cascades come to the front. Environmental change, deforestation, and uncontrolled advancement undermine the fragile equilibrium that supports these regular marvels. The hazy cloak that hides the cascades turns into an illustration for the vulnerability that weavers their future.

In the midst of these difficulties, in any case, lies the flexibility of nature and the human soul. Preservation endeavors, local area drives, and the aggregate will to safeguard these fountains arise as encouraging signs. The account digs into the systems utilized to save the immaculate magnificence of India's cascades, guaranteeing that they keep on murmuring their accounts for a long time into the future.

In the finishing up sections, the story turns its look toward what's to come. What lies ahead for India's fountains in a world set apart by fast natural changes? The conversation incorporates the job of dependable the travel industry, feasible practices, and the basic for an agreeable concurrence among mankind and nature. The closing comments reverberation a confident feeling, recognizing that the fountains, with their immortal stories, will continue as images of elegance and magnificence in the developing story of the Indian subcontinent.

Basically, the prologue to "India's Beauty: Fountains Murmuring Stories" is a tribute to the charming universe of cascades that embellish the subcontinent. It is an encouragement to investigate the geographical wonders, social accounts, and ecological difficulties that shape these fountains into living narratives. As the excursion unfurls, the peruser is encouraged to listen near the murmurs of the falling waters, for inside them lie the enthralling accounts of India's superb fountains.

2. **Explore the cultural significance of waterfalls in Indian mythology and history.**

The social meaning of cascades in the embroidery of Indian folklore and history uncovers a rich and complex story that rises above the simple actual presence of these fountains. In the huge and various scene of the Indian subcontinent, cascades have not exclusively been topographical developments however have likewise taken on profound, representative, and verifiable aspects, meshing themselves into the actual texture of the way of life.

In the broad material of Indian folklore, cascades are many times depicted as hallowed destinations, accepted to be the habitation of divinities or the sign of heavenly energy. The Ganges, starting from the Gangotri Glacial mass in the Himalayas, is a perfect representation. The plunge of the Ganges from the sky to the Earth, worked with by Ruler Shiva's demonstration of getting the waterway in his locks, is a legendary story implanted in Hindu conviction.

The various cascades along the course of the Ganges, like the Devprayag and

Rudraprayag, are loved as critical journey locales, and taking a plunge in their waters is viewed as sanitizing for the spirit.

Past the Himalayas, the Western Ghats likewise harbor cascades that have cut a specialty in folklore. Run Falls, for example, is related with the unbelievable figure of Rani Keladi Chennamma, a sovereign who took shelter in the Run Falls region during a conflict. The social meaning of this outpouring is intertwined with stories of valiance and flexibility, making an association between the normal scene and verifiable occasions.

The social embroidered artwork is additionally enhanced by the conviction that cascades act as entryways to different domains. In numerous native customs, these fountains are seen as entryways between the natural domain and the heavenly, making a feeling of wonderment and worship. The fog that ascents from the falling waters frequently turns into a representation for the shroud between the seen and the concealed, welcoming travelers and fans to participate in an otherworldly excursion.

Cascades in Indian folklore are not only actual elements; they are living epitomes of the consecrated. The flowing waters are viewed as purifiers, washing away pollutions and sins. Pioneers embrace strenuous excursions to arrive at these hallowed fountains, accepting that the demonstration of washing in their waters will scrub the spirit and prepare for otherworldly illumination. The social ceremonies related with cascades become an indispensable piece of the cultural and strict works on, representing the timeless pattern of life, demise, and resurrection.

The authentic meaning of cascades in India is entwined with the civilizations that thrived along their banks. Antiquated urban communities and settlements frequently created close to these water sources, using the regular overflow for food and development. The Bhedaghat Marble Rocks on the Narmada Waterway, decorated with Dhuandhar Falls, stand as a demonstration of this verifiable incorporation of cascades into human progress. The stones, etched by the flowing waters over hundreds of years, feature the agreeable concurrence among nature and human inventiveness.

Cascades have likewise assumed an essential part in the protection of old strongholds. The Run Falls, settled in the Western Ghats, filled in as a characteristic obstruction, making it hard for trespassers to break the strongholds. This verifiable setting adds layers of importance to the cascades, as they become images of normal magnificence as well as essential components in the embroidery of fighting and guard.

Additionally, the portrayal of cascades in old craftsmanship and writing highlights their social significance. From antiquated texts like the Vedas to old style Sanskrit verse, cascades find wonderful articulation that rises above their actual qualities.

The musical plunge of the waters turns into a similitude for the repeating idea of life, mirroring the rhythmic movement of human encounters.

The approach of different administrations and rulers in Indian history has additionally transformed the social meaning of cascades. The leaders of archaic India, like the Cholas and the Hoysalas, commended the normal excellence of their spaces, including the cascades that enhanced their regions. Sanctuaries and landmarks were many times developed in vicinity to these fountains, coordinating the normal and the heavenly in the design scene.

As the account of India's social woven artwork unfurls, it becomes clear that cascades are not simple geological highlights; they are storehouses of stories, legends, and authentic occasions. They are loved as appearances of the heavenly, wellsprings of profound virtue, and observers to the development of human progress. The social meaning of cascades in Indian folklore and history is a demonstration of the significant interconnection between nature, otherworldliness, and the human experience.

3. **Set the stage for a journey through the diverse landscapes where these waterfalls grace the land.**

The stage is set, and the drapery ascends on an excursion through the different and charming scenes of India, where the smooth presence of cascades paints a clear embroidery across the subcontinent. As we leave on this odyssey, the kaleidoscope of geological miracles unfurls before our eyes, welcoming us to cross mountains, valleys, levels, and woodlands where the fountains murmur stories as antiquated as time itself.

In the Western Ghats, a mountain range that stretches along the western edge of the Indian landmass, the excursion starts. The Ghats, frequently alluded to as the "Incomparable Slope of India," are a geographical wonder that supports a surprising cluster of biodiversity. Here, the scene is embellished with verdant slopes, thick timberlands, and, obviously, the glorious cascades that fountain down with both power and effortlessness. As the excursion takes us more profound into this locale, famous cascades, for example, Run Falls, arranged in the Shimoga area of Karnataka, reveal their magnificence.

Run Falls, the second-most elevated plunge cascade in India, is a demonstration of the glory of nature's plan. The Sharavathi Stream, plunging from the Western Ghats, takes a tremendous jump of around 830 feet, making a stunning outpouring that spellbinds all who see it. The thundering waters dive into the foggy chasm underneath, making a visual display that represents the spectacular force of nature. Here, the air is loaded down with the strengthening fragrance of wet earth, and the flowing waters turns into an ensemble that resounds through the encompassing scene.

As we adventure further into the Western Ghats, Athirapally Falls in Kerala divulges its own story. Settled in the midst of the lavish plant life of the Sholayar

Timberlands, Athirapally is frequently alluded to as the "Niagara of India." The Chalakudy Stream, plunging from the Anamudi Mountains, makes an outpouring that traverses more than 330 feet in width. The fog rises like ethereal shades, covering the scene in a secretive charm. Here, the cascade isn't simply something else; it is a living element, a power of nature that shapes the biology of the district and rouses stunningness in the people who stand observer to its highness.

Dudhsagar Falls, situated in the verdant scenes of Goa, adds one more layer to the story of the Western Ghats. Deciphered as the "Ocean of Milk," Dudhsagar is a multi layered cascade that fountains from a level of north of 1,000 feet. The Mandovi Waterway, starting in the Western Ghats, makes a visual display as its waters dive into the tough landscape underneath. The smooth white appearance of the falling waters, as though pouring from the sky, adds a legendary appeal to the scene.

Moving toward the north, the excursion climbs to the lofty Himalayan lower regions, where the scene assumes an alternate personality. Here, cascades are land developments as well as are unpredictably woven into the profound and social embroidered artwork of the locale. The consecrated Ganges, beginning from the Gangotri Ice sheet, plummets from the levels of the Himalayas, making various cascades along its course.

The otherworldly meaning of these cascades is significant. Devprayag, where the Bhagirathi and Alaknanda Streams join to shape the Ganges, is viewed as a consecrated conversion. The wild waters, as they participate in a heavenly association, become an image of immaculateness and heavenly energy. The fountains at Devprayag are not only actual peculiarities; they are the encapsulation of the hallowed waterway's excursion from the sky to the natural domain.

Rudraprayag, one more sacrosanct site along the Ganges, is named after Master Rudra (Shiva). The fountains here are accepted to repeat the grandiose dance of Ruler Shiva, and travelers run to this site to observe the heavenly indication. The entwining of folklore and geology in the Himalayas changes the cascades into living exemplifications of otherworldly energy.

As the excursion unfurls, the investigation stretches out toward the eastern locales of India, where lesser-investigated cascades uncover themselves as unlikely treasures. Here, the scenes are portrayed by undulating slopes, thick backwoods, and wandering streams. The cascades in the eastern domains are normal marvels as well as essential components of the networks that live in their closeness.

The fountains in the East become piece of the daily existences of individuals, filling in as wellsprings of food, water system, and social importance.

In Meghalaya, the "Homestead of Mists," cascades, for example, Nohkalikai Falls overflow down from raised levels, making an entrancing display. Nohkalikai, the most elevated plunge cascade in India, is encircled by thick vegetation, adding a component of persona to its presence. Here, the cascades are not far off ponders; they are buddies in the cadence of life.

The southern areas of India, especially the Deccan Level, contribute their own cadence to the story. The scene changes once more, uncovering an embroidery of rough landscape, rich fields, and thick timberlands. The cascades in the Deccan Level, whether in the Western Ghats or the Eastern Ghats, deliver an alternate tasteful.

In the Western Ghats, the Run Falls, investigated prior in the excursion, represent the magnificence of the Deccan Level. The musical drop of the Sharavathi Waterway makes an outpouring that reverberates with the exceptional energy of the locale. In the Eastern Ghats, the Kiliyur Falls in the Shevaroy Slopes of Tamil Nadu offer a more close encounter. Encircled by thick vegetation, the falls make a peaceful environment, welcoming reflection and thought.

However, in the midst of the magnificence and loftiness of these cascades, challenges loom not too far off. The excursion takes an impactful turn as the story defies the natural issues looked by India's cascades. Environmental change, deforestation, and uncontrolled improvement compromise the fragile equilibrium that supports these normal miracles. The fog that ascents from the falling waters turns into a representation for the vulnerability that weavers their future.

In any case, inside these difficulties lies the versatility of nature and the human soul. Preservation endeavors, local area drives, and the aggregate will to safeguard these fountains arise as encouraging signs. The account digs into the techniques utilized to save the immaculate excellence of India's cascades, guaranteeing that they keep on murmuring their accounts for a long time into the future.

In the closing parts, the story turns its look toward what's to come. What lies ahead for India's fountains in a world set apart by quick ecological changes? The conversation includes the job of capable the travel industry, feasible practices, and the basic for an agreeable conjunction among humankind and nature.

As we arrive at the finish of this excursion through India's different scenes, where cascades beauty the land, the peruser is left with a significant appreciation for the interconnectedness of topography, culture, and the regular world. The fountains, with their immortal stories, persevere as images of elegance and magnificence in the developing story of the Indian subcontinent. The stage, when set, turns into an immortal background against which the cascades keep on murmuring their captivating stories.

Chapter 1

Nature's Symphony Begins

In the tranquil first light of another day, Nature's ensemble starts, organized by the unpretentious developments of the components. As the main light breaks into the great beyond, painting the sky with tones of pink and gold, an orchestra of varieties reports the enlivening of the world. The sun, a brilliant guide, coordinates its beams across the scene, reinvigorating the lethargic earth. The air conveys a fresh newness, as though Nature itself is breathing in a full breath, prepared to breathe out the songs of another day.

The orchestra unfurls with the delicate stirring of leaves, a preface to the amicable dance of the breeze through the trees. Each leaf turns into a note, making a song that resounds with the cadence of the morning breeze. The branches influence as one, their developments arranged by the concealed hands of nature. An artful dance rises above human comprehension, a dance that has been performed since days of yore.

In the midst of this arboreal expressive dance, the avian troupe participates. Birds of horde feathers become the entertainers, their melodies winding through the air like a sensitive string sewing the ensemble together.

From the bright twittering of robins to the eerie songs of grieving pigeons, each bird adds a special note to the many-sided sythesis. Their calls reverberation through the tremendous breadth, changing the environmental factors into a church of sound, where each animal is an admirer in today custom.

The water components contribute their own stanzas to the ensemble. Streams prattle and waterways rush, their ebbs and flows making a liquid congruity that goes with the general song of the breeze and birdsong. Nature's fluid notes overflow over rocks, their musical examples adding a percussive component to the structure. The delicate lapping of waves against the shore turns into a calming refrain, a sign of the interconnectedness of all components in this stupendous organization.

As the day advances, the orchestra develops. The sun moves higher overhead, giving occasion to feel qualms about its warm gleam the scene. Nature's tones become more energetic, and the soundscape adjusts to the evolving rhythm. In the glades, the stirring grass turns into a murmured percussion, the influencing blossoms a visual portrayal of the continuous crescendo. Butterflies and honey bees, the fragile artists of the day, shudder from one sprout to another, their wings making an unpretentious murmur that fits with the continuous tune.

Amidst this ensemble, the human presence turns into a member instead of a simple eyewitness. Strides on the timberland floor, the giggling of youngsters, and the mumble of discussions mix flawlessly into the normal soundscape. Nature benevolently obliges the human component, coordinating it into the all-encompassing structure. It is an update that people, as well, are necessary to the many-sided trap of life, contributing their own notes to the ensemble of presence.

As the sun starts its plunge, the orchestra goes through an inconspicuous change. The diurnal animals step by step yield the stage to their nighttime partners. The stirring leaves and avian tunes clear a path for the nighttime sounds — the hooting of owls, the croaking of frogs, and the far off crying of wolves. The night sky, enhanced with stars, turns into an enormous background to this heavenly ensemble.

In the quietness of the evening, the ensemble takes on a more thoughtful tone. The stirring leaves presently murmur mysteries of the day's occasions, and the nighttime animals take part in an exchange that rises above the limits of species. It is a period of reflection, a melodic break before the orchestra resumes with the main light of sunrise.

Nature's orchestra is a persistent, consistently changing show-stopper that unfurls in amicable waves from dawn to dusk and then some. It is a demonstration of the mind boggling equilibrium of environments, the interconnectedness of every living being, and the recurrent idea of life. To drench oneself in this ensemble is to turn into a piece of the bigger organization, a modest note in the stupendous score composed by the universe.

1.1 Delve into the geological formations that give rise to India's stunning waterfalls.

Digging into the geographical developments that lead to India's dazzling cascades reveals an enamoring story of nature's chiseling ability. From the northern compasses of the Himalayas toward the southern levels, India's geography is a material whereupon cascades paint their stunning representations. These fountains are not simply transient scenes; they are land narrators, murmuring stories scratched in the actual bedrock of the subcontinent.

The great Himalayas, standing tall in the north, contribute essentially to India's cascade woven artwork. As waterways beginning from the icy masses overflow down the tough slants, they experience a different cluster of rock developments. The geographical piece of the Himalayan locale, portrayed by a blend of sedimentary and transformative rocks, makes a powerful scene for the development of cascades. The powerful plummet of these streams, affected by the precarious slopes and fluctuated rock types, brings about the absolute most fabulous cascades in the country.

In the northeastern piece of India, the Seven Sisters Falls represent the geographical show of the Himalayan lower regions. Every one of the seven individual cascades, named after the northeastern states, is a demonstration of the unpredictable exchange of rocks and water. Here, the streams explore through steep crevasses, experiencing safe stone layers that block their stream. At the point when the water finds a way through the gentler stone, it overflows down in a progression of stunning dives, making a visual ensemble of fog and movement.

Moving towards the focal locale of India, the Deccan Level unfurls its own geographical account. This tremendous scope of raised territory, dominatingly made out of basaltic stone, fills in as the material for cascades that manifest the district's volcanic heritage. The magma streams from old volcanic action have abandoned a geology portrayed by basalt segments and ventured developments. Cascades like the Run Falls in Karnataka embody the marriage of land processes, where the waterways, having dissolved through the basalt, make a flowing exhibition over the tough strides of the level.

The Western Ghats, an UNESCO World Legacy site, further decorate India's cascade collection. This mountain range, extending lined up with the western coast, is a gold mine of biodiversity and geographical marvels. The Ghats' mind boggling topographical history, set apart by structural exercises and enduring cycles, makes way for the production of various cascades. Gleaming drapes of water plunge from the bluffs, etched by the incessant interchange of enduring, disintegration, and the basic stone designs.

In the southern piece of the Western Ghats, the Athirappilly Falls in Kerala exhibit the land theater at its best. Here, the Chalakudy Stream explores through a rough territory, experiencing groups of laterite and gneiss. The water, in its determined excursion, cuts through the stone, chiseling a progression of fountains that finish in a stunning dive. The land variety of the locale, described by old Precambrian rocks, adds to the remarkable person of every cascade.

The Eastern Ghats, however not too known for their cascades, assume a urgent part in India's hydrological show. The fluctuated land arrangements around here, including stones, schists, and gneisses, add to the production of cascades that frequently stay unlikely treasures. The Kiliyur Falls in Tamil Nadu, for example, overflow down the Shevaroy Slopes, uncovering the topographical wealth of the Eastern Ghats.

Past the levels and mountain goes, the alluvial fields and waterfront districts of India harbor their own cascade stories. The Western waterfront district, with its lateritic soil and moving slopes, is embellished with cascades like the Dudhsagar Falls in Goa. As the Mandovi Waterway dives from the raised level, the fountain wanders through rich plant life, uncovering the topographical show that shapes the waterfront scenes.

The geographical developments adding to India's cascades are not static; they are dynamic elements going through steady change. The powers of enduring, disintegration, and structural movement keep on trim the scene, making new accounts for the cascades to express. The stones that guide the progression of water have seen ages unfurl, and in their quiet strength, they portray the topographical adventure of the subcontinent.

The land developments molding India's cascades are an enamoring declaration to the World's consistently evolving material. As we dive further into this land story, it becomes clear that the development of cascades is a powerful transaction of different variables, each adding to the display in its own particular manner.

The enduring cycles, driven by components like breeze, water, and temperature vacillations, are instrumental in forming the stones that guide the course of waterways and streams. In the Himalayas, for instance, the steady powers of enduring follow up on the assorted stone sorts, separating them into dregs that structure the riverbeds. The Ganges and its feeders, having crossed through the tough mountain landscape, convey with them the dissolved leftovers of antiquated rocks. This residue loaded water, as it slips from the levels, turns into a stone carver, cutting through the scene and bringing forth cascades.

Disintegration, a nearby ally to enduring, emphasizes the land show. The constant progression of water over the stones slowly erodes the surfaces, making mind boggling examples and uncovering the basic designs. In the Deccan Level, the consistent erosive activity of streams like the Sharavathi has prompted the arrangement of profound canyons and valleys.

The Run Falls, flowing over the rough strides of the level, is a demonstration of the erosive force of water and job in molding the geographical elements characterize the scene.

Structural action, a topographical power of significant importance, assumes an essential part in the production of cascades. The Indian subcontinent, situated at the united limit of the Indian and Eurasian structural plates, encounters continuous structural developments. The Himalayas, the consequence of the impact between these plates, are a demonstration of the groundbreaking force of structural powers. The elevating of the Himalayan reach not just makes way for the streams' drop yet additionally impacts the geographical qualities of the stones they experience.

Volcanic action, another geographical power, influences the scene and adds to the arrangement of cascades. The Deccan Level, with its basaltic stone arrangements, validates the old volcanic emissions that molded the district. The magma streams, hardened over the long run, make a hearty starting point for waterways to shape their ways. The basalt segments at places like the Run Falls stand as quiet observers to the volcanic heritage that characterizes the land personality of the level.

As streams navigate different land territories, they experience changing stone sorts that direct the idea of the cascades. Harder, safe rocks obstruct the progression of water, prompting the making of steep drops and plunges. Milder rocks, then again, capitulate to disintegration all the more promptly, bringing about gentler fountains. The geographical heterogeneity of the Western Ghats, for example, adds to the variety of cascades that decorate the locale. The layered stone arrangements and differed pieces become the range for nature's imaginative articulation.

The topographical time scale unfurls in the stratigraphy of rocks, uncovering the tales of old scenes and climatic circumstances. The Precambrian rocks of the Western Ghats, going back large number of years, demonstrate the veracity of the topographical history of the subcontinent. The layers of sedimentary rocks in the eastern locales describe stories of old oceans and the sluggish cycles that formed the geology. Cascades, in their endless plunge, become people who goes back and forth through time, conveying with them the reverberations of the land ages they navigate.

The job of human exercises in molding the topographical setting of cascades can't be overlooked. Deforestation, urbanization, and industrialization modify the regular equilibrium, impacting the residue load in waterways and the security of rock arrangements. Human intercessions, once in a while unexpected, can speed up disintegration or disturb the sensitive balance that supports cascades. The sensitive environment encompassing these fountains, with its vegetation, is complicatedly connected to the land setting, and any aggravation can have flowing impacts.

In understanding the land developments of India's cascades, perceiving the interconnectedness of Earth's processes is vital. The hydrological cycle, driven by the sun's energy, arranges the development of water across the scene. The geographical arrangements go about as channels, directing this stream and forming the fountains that dazzle the human creative mind. A comprehensive viewpoint rises above individual cascades, enveloping the more extensive elements of the World's topography and the complicated connections between different components.

The geographical developments that lead to India's shocking cascades are an orchestra of powers, a joint effort between the World's components that unfurls over land ages. From the transcending levels of the Himalayas to the undulating levels and seaside districts, each scene recounts to a remarkable story carved in the stone and water. India's cascades, with their entrancing magnificence, act as topographical narrators, describing stories of enduring, disintegration, structural developments, and the determined section of time. To dig into the geographical developments of these fountains is to set out on an excursion through the chronicles of Earth's set of experiences, where shakes and water meet to make scenes that reverberation the significant excellence of nature's land orchestra.

1.2 Showcase iconic cascades such as Jog Falls, Athirapally Falls, and Dudhsagar Falls.

Venturing through the different scenes of India, one experiences famous fountains that stand as demonstration of the country's normal wonder. Among these, Run Falls, Athirapally Falls, and Dudhsagar Falls arise as praised delegates of India's rich cascade embroidery, each having a particular appeal formed by its one of a kind geological setting.

Run Falls, settled in the rough landscape of the Western Ghats, is a superb exhibition that encapsulates the glory of nature's powers. Shaped by the Sharavathi Waterway, Run Falls remains as the second-most elevated plunge cascade in India. As the waterway makes its plummet from the Deccan Level, it experiences a progression of rough precipices made out of basalt rock. The land show unfurls as the waterway overflows over the slope, falling a stunning 830 feet in a sheer upward drop.

The appeal of Run Falls lies in its level as well as in the four unmistakable fountains that comprise its loftiness. Raja, Rani, Meanderer, and Rocket, the four portions, make a visual orchestra as they stream over the slope. During the rainstorm season, Run Falls changes into a thundering downpour, a demonstration of the colossal volume of water that plunges over the bluffs. The air becomes immersed with fog, and the flowing water resounds through the valley, making a remarkable encounter for those lucky enough to observe this regular exhibition.

The geographical elements encompassing Run Falls assume an essential part in forming its personality. The basaltic stone developments, leftovers of old volcanic movement in the Deccan Level, give a durable establishment to the fountains. The columnar jointing in the basalt rock, described by hexagonal crystals, adds to the visual allure of the falls. These regular hexagonal developments, a consequence of the cooling and cementing of magma, make a hypnotizing design that improves the land feel of Run Falls.

In the southern spans of the Western Ghats, Athirapally Falls enthralls with its captivating magnificence. Arranged on the Chalakudy Waterway in Kerala, this cascade is frequently alluded to as the "Niagara of India." The Chalakudy Stream, exploring through the undulating territory of the Western Ghats, experiences a progression of rough outcrops and steep precipices. As the waterway plunges over these slopes, Athirapally Falls comes to fruition, flowing over an expansiveness of 330 feet.

The lavish plant life that encompasses Athirapally Falls adds to its charm, making a beautiful setting that supplements the crude force of the flowing water. The encompassing Sholayar Reach, shrouded in thick timberlands, adds to the persona of the scene, making Athirapally Falls a visual pleasure as well as a vivid involvement with the core of nature. The falls act as a dynamic environment, with a rich variety of verdure flourishing in the dampness loaded air and prolific soils.

Geographically, the locale adds to the particular person of Athirapally Falls. The stones in the Western Ghats, essentially made out of antiquated Precambrian arrangements, impact the course of the waterway and the making of cascades. The presence of gneiss and laterite rock developments in the district assumes a part in molding the fountains. The topographical history implanted in these stones turns into a basic piece of the account told by Athirapally Falls, a story scratched in the World's outside north of millions of years.

Dudhsagar Falls, arranged in the thick Western Ghats of Goa, is a visual event that enamors with its sheer scale and beautiful magnificence. The Mandovi Stream, wandering through the undulating scenes of the Western Ghats, experiences a steep drop in the rise. Dudhsagar, meaning "Ocean of Milk," suitably portrays the presence of the falls as the water overflows over a transcending precipice, making a foamy and ethereal exhibition.

Dudhsagar Falls is special not just for its level, remaining at an amazing 1,017 feet, yet in addition for its layered design. The fountains structure a progression of four unmistakable levels, each adding to the general grandness of the falls. The encompassing scene, described by rich plant life and dynamic biodiversity, adds to the vivid experience of Dudhsagar Falls. The Bhagwan Mahavir Untamed life Safe-haven, through which the falls are available, further improves the normal appeal of the district.

Geographically, Dudhsagar Falls mirrors the impact of the Deccan Trap basalts in the Western Ghats. The level's volcanic history, set apart by broad magma streams and basaltic developments, gives an establishment to the production of cascades. The layered construction of the basaltic stone adds to the layered idea of Dudhsagar Falls, displaying the geographical complexities of the area. The falls, encompassed by thick woodlands and overflowing with biodiversity, become a microcosm of the Western Ghats' environmental extravagance.

The land stories of Run Falls, Athirapally Falls, and Dudhsagar Falls feature the assorted cycles that shape India's famous fountains. The Western Ghats, with its shifted land developments, remains as an ongoing idea interfacing these cascades. The Deccan Level, reaching out across focal India, contributes its own geographical person to the scene. The mind boggling dance among rock and water, molded by structural powers, enduring, and disintegration, unfurls in each outpouring, making a topographical orchestra that reverberations through time.

Run Falls, Athirapally Falls, and Dudhsagar Falls, however topographically particular, share normal components that characterize their land personality. Basaltic stone developments, a tradition of old volcanic movement, assume a critical part in forming these cascades. The columnar jointing in basalt, as found in Run Falls, adds a visual aspect to the land story. The Western Ghats, with its blend of gneiss, laterite, and old Precambrian rocks, gives a different geographical range to the making of fountains like Athirapally Falls and Dudhsagar Falls.

Past their topographical importance, these cascades become social and biological touchpoints. They are loved in nearby fables, celebrated in writing, and act as imperative biological systems supporting different verdure. The encompassing scenes, molded by the land developments, become safe-havens for biodiversity and centers of normal excellence.

Proceeding with our investigation of Run Falls, Athirapally Falls, and Dudhsagar Falls, it is fundamental to dig into the social and biological importance implanted in these notable fountains. Past their geographical stories, these cascades arise as living biological systems, social images, and crucial supporters of the different scenes they effortlessness.

Run Falls, with its sensational loftiness, isn't only a geographical wonder yet a social symbol in the district. The fountains, referred to locally as Geruoppe Falls, hold a critical spot in the old stories and customs of Karnataka. The thundering waters of Run Falls are compared to the mane of a pony, moving the name "Run," and that signifies "hair" in Kannada. This social affiliation changes the falls into an image of solidarity and essentialness, a demonstration of the consistent coordination of nature into the social texture of the district.

Enthusiasts frequently visit Run Falls as a component of their journey to the close by Shringeri Sharadamba Sanctuary. The vicinity of the tumbles to strict locales adds a profound aspect to their social importance. Pioneers, while looking for divine gifts, wind up within the sight of the strong fountains, supporting the interconnectedness of nature and otherworldliness.

The social pertinence of Run Falls stretches out to its portrayal in writing, craftsmanship, and provincial merriments. Artists and essayists have deified the falls in their stanzas, catching the pith of its power and magnificence. The falls likewise become a point of convergence during celebrations, where the encompassing networks meet up to commend the lively social embroidery of the locale. Run Falls, with its immortal presence, in this way turns into a social reference point that rises above its topographical beginnings.

Athirapally Falls, settled in the verdant scenes of Kerala, also holds social importance well established in the locale's practices. The falls, frequently alluded to as the "Niagara of India," have tracked down a spot in the realistic world, highlighting in various Indian movies. This social openness through film has raised Athirapally Tumbles to famous status, drawing vacationers and nature devotees from the nation over.

The Athirapally district, encompassed by lavish timberlands, is occupied by native networks whose lives are complicatedly woven into the regular embroidered artwork of the scene. The falls, with their profound appeal, are viewed as hallowed by the neighborhood networks. The ethereal excellence of Athirapally turns into a wellspring of motivation for specialists and journalists, molding the social story of the locale.

Naturally, Athirapally Falls fills in as a biodiversity area of interest. The Sholayar Reach, encompassing the falls, is home to a rich assortment of greenery. The clammy environment produced by the falls makes a microenvironment helpful for the flourishing of different plant species. The encompassing backwoods harbor endemic and imperiled species, making the Athirapally area socially critical as well as biologically fundamental.

Dudhsagar Falls, arranged in the midst of the rich woodlands of Goa, conveys its own social and biological stories. The name "Dudhsagar," signifying "Ocean of Milk" in Hindi, mirrors the frothy appearance of the flowing water as it drops from the transcending precipices. Neighborhood old stories credits the tumbles to the requests of an in the princess milk-like waters, underscoring the social combination of fantasy and regular magnificence.

The environmental significance of Dudhsagar Falls is highlighted by its area inside the Bhagwan Mahavir Natural life Asylum. The falls act as a help for the different cluster of greenery possessing the safe-haven. The wet climate made by the water splash supports an assortment of plant animal categories, while the encompassing woodlands give a living space to various creature species, including elephants, deer, and extraordinary birds.

The Mandovi Waterway, which brings forth Dudhsagar Falls, likewise assumes a critical part in the social and financial existence of the district.

The waterway's waters are bridled for different purposes, from horticulture to hydroelectric power age. This human instinct association mirrors the complicated harmony between social practices and environmental maintainability, highlighting the reliance of networks on the regular assets given by the scene.

As we mull over the social and environmental elements of these cascades, it becomes clear that they are not confined peculiarities but rather essential parts of the more extensive biological systems and social scenes they occupy. Individuals living in closeness to these fountains are many times stewards of their regular legacy, and the cascades, thusly, shape the social practices and characters of these networks.

Run Falls, Athirapally Falls, and Dudhsagar Falls, with their topographical loftiness, social imagery, and environmental essentialness, stand as representatives of the sensitive harmony between human exercises and the regular world. They coax us to see the value in the significant associations between land processes, social stories, and the food of biodiversity. These cascades are not simply static elements in the scene; they are dynamic substances that wind around together the strings of topography, culture, and biology into an amicable embroidery.

The exhibit of Run Falls, Athirapally Falls, and Dudhsagar Falls uncovers a multi-layered story that stretches out past the geographical developments. These fountains encapsulate the social characters of their districts, going about as respected images and wellsprings of motivation. At the same time, they capability as indispensable environments, supporting different plant and creature life. The social and biological elements of these cascades highlight the perplexing connections between human social orders and the regular habitats they possess, underlining the requirement for a comprehensive way to deal with preservation and enthusiasm for our planet's normal miracles.

1.3 Share anecdotes and local legends associated with these natural wonders.

Investigating the normal marvels of Run Falls, Athirapally Falls, and Dudhsagar Falls reveals an embroidery of tales and neighborhood legends that add a layer of charm to these generally stunning scenes. These stories, went down through ages, entwine with the land, social, and environmental aspects, making a rich story that reverberates with the soul of the locales they elegance.

Run Falls, with its roaring fountains and sensational precipices, isn't just a geographical wonder yet in addition a material for nearby fables. One of the persevering through stories related with Run Falls is established in the social texture of the Karnataka locale.

Privately known as "Geruoppe Falls," the falls are said to get their name from the Kannada words "Geru" (signifying "thunder") and "Hoppe" (signifying "fall"). This classification highlights the loud solid produced by the plunging waters, repeating like the thunder of a strong animal.

Notwithstanding the etymological historical underpinnings, Run Falls is attached to an entrancing legend including a neighborhood god. As indicated by prevalent thinking, the Sharavathi Waterway, from which Run Falls starts, is viewed as the natural sign of the goddess Sharavathi. The falls, then, at that point, are viewed as the goddess herself plummeting to favor the land. Explorers and local people the same worship Run Falls for its land superbness as well as for the otherworldly association they feel with the heavenly presence encapsulated in the flowing waters.

Likewise with numerous regular marvels, Run Falls has motivated artists and scholars to mesh its highness into stanzas and exposition. The falls have turned into a dream for imaginative articulation, with writers endeavoring to catch the incredible excellence and power through words. This social reverberation further extends the profound association that individuals feel with Run Falls, raising it past a geographical scene to a wellspring of imaginative motivation and otherworldly reflection.

Athirapally Falls, arranged in the midst of the rich scenes of Kerala, isn't just eminent for its realistic appeal yet additionally for the nearby legends that cover its flowing magnificence. One such legend is established in the fables of the district and includes a legendary romantic tale. It is said that the stream spirits, enchanted by the stunning magnificence of the falls, frequently assemble here to observe the timeless dance of affection between the water and the stones.

The Athirapally district, being wealthy in biodiversity and encircled by thick backwoods, is accepted to be a gathering point for divine creatures. The falls, with their ethereal excellence, act as a heavenly amphitheater where divine elements gather. Neighborhood people group share tales of seeing secretive lights and sounds close to the falls, ascribing them to the concealed presence of these divine creatures. This magical perspective adds a layer of charm to Athirapally Falls, making it something other than an actual display however where the supernatural and regular universes join.

Athirapally Falls likewise tracks down a spot in neighborhood ancestral legends. The Kadar clan, native to the area, has stories that interlace with the falls and the encompassing Sholayar Reach. As per ancestral stories, the falls are viewed as holy, and the spirits dwelling in the woods around Athirapally are accepted to safeguard the normal equilibrium of the district. The falls, in this story, become a gathering point between the natural and heavenly domains, where the spirits guarantee the concordance of nature.

The realistic unmistakable quality of Athirapally Falls has additionally imbued its picture in the social awareness. The falls filled in as a setting for various Indian movies, enhancing its prevalence and drawing in guests from all over. The effect of film on the social meaning of regular tourist spots is clear in the manner individuals see and draw in with them, making a scaffold among reel and genuine scenes.

Dudhsagar Falls, meaning the "Ocean of Milk," in Goa is wrapped in old stories that adds a dash of persona to its transcending magnificence. One of the neighborhood legends portrays the tale of a princess who used to wash in the unblemished waters of Dudhsagar. It is said that she would pour a container of sweet milk into the falls prior to venturing into the water. The milk, flowing down with the falling water, changed into a foamy, white mass, making the deception of an ocean of milk. This captivating story loans the falls its name as well as adds to the supernatural emanation that encompasses Dudhsagar.

The legend of Dudhsagar Falls reaches out to its relationship with Master Krishna. As indicated by Hindu folklore, Master Krishna is frequently portrayed with a container of milk, and the foamy white waters of Dudhsagar are accepted to represent this heavenly contribution. Travelers visiting the close by Dudhsagar sanctuary frequently drench themselves in the waters, thinking of it as a holy demonstration that associates them with the heavenly pith typified in the falls.

Past these conventional stories, Dudhsagar Falls is likewise entwined with the later history of the district. During the pilgrim period, the falls filled in as an essential area for the Goa-Vasco railroad course. The train tracks, passing by the falls, add a component of mankind's set of experiences to the normal miracle. The leftovers of the railroad span, presently unused, stand as quiet observers to the progression of time and the changing scenes of Goa.

Similarly as with different falls, Dudhsagar is profoundly implanted in the social acts of the neighborhood networks. The Mahavir Untamed life Safe-haven, where the falls are arranged, isn't just a shelter for biodiversity yet in addition a holy space for the ancestral networks. The love of gods related with the normal components, including water, is a typical social practice among these networks. Dudhsagar Falls, as a sign of the heavenly in nature, turns into a fundamental piece of these social customs and festivities.

These tales and neighborhood legends related with Run Falls, Athirapally Falls, and Dudhsagar Falls give a focal point through which we can see these normal miracles. They become something beyond land developments; they are living substances with stories that reverberate with the social and profound personalities of the districts they enhance. The oral customs went down through ages act as a demonstration of the persevering through association among networks and the scenes they possess.

Diving further into the captivating stories and nearby legends encompassing Run Falls, Athirapally Falls, and Dudhsagar Falls uncovers an embroidery woven with social subtleties, otherworldly convictions, and a profound veneration for the regular world. These accounts intensify the appeal of these cascades as well as mirror the significant associations between human social orders and the scenes they occupy.

Run Falls, with its loud thunder and forcing precipices, holds a unique spot in the hearts of individuals of Karnataka. Past its land brilliance, the falls are complicatedly associated with territorial celebrations and ceremonies. During the promising month of Shravana in the Hindu schedule, Run Falls observers a flood of travelers who set out on an excursion to give proper respect to the Sharavathi Stream, thought about a sacrosanct sign of the goddess Sharavathi.

The strict meaning of Run Falls is increased during the celebration of Bother Panchami when individuals come to look for gifts and perform ceremonies at the close by Kollur Mookambika Sanctuary. The falls, settled in the lap of the Western Ghats, become a profound safe house where the combination of nature and holiness is commended. Explorers frequently take part in ceremonies that include offering coconuts and other representative things into the waterway, offering their thanks and respect for the sacrosanct waters.

Neighborhood celebrations additionally integrate Run Falls into their social stories. The falls act as a background for customary dance and music exhibitions, improving the merry climate with the normal glory of the district. The reverberation of these social festivals enhances the connections between the neighborhood networks and Run Falls, making it a necessary piece of their aggregate character.

The legends encompassing Run Falls grows to incorporate stories of neighborhood legends and legendary characters. Legends portray the endeavors of striking people who explored the misleading precipices and tempestuous waters of the falls. These accounts, went down through ages, commend the unyielding human soul and the cooperative connection among individuals and their regular environmental elements.

Athirapally Falls, with its rich environmental factors and artistic appeal, is saturated with legends that reverberation through the Western Ghats of Kerala. The falls, alluded to as "Punnagai Mannan Cascades" after a popular Tamil film shot there, have turned into a seal of normal magnificence in the social cognizance. Nonetheless, the neighborhood legends add a layer of persona to the fountains.

One such legend includes the antiquated ancestral networks of the district. As per ancestral legends, the spirits dwelling in the Sholayar Reach are accepted to be watchmen of the Athirapally Falls.

These generous spirits, known as "Kumara Yakshi" and "Kumara Yakshi," are said to safeguard the falls and guarantee the prosperity of the encompassing timberlands. The stories of these watchman spirits interweave with the biological personality of Athirapally Falls, depicting an amicable connection between the otherworldly and regular domains.

Athirapally Falls likewise tracks down a spot in the social articulations of the native networks. Conventional dance structures, for example, Theyyam and Kathakali, frequently integrate the falls into their exhibitions, involving the normal scene as a phase for narrating. These imaginative articulations not just exhibit the inventive splendor of the nearby networks yet additionally act as a mode for safeguarding and communicating social stories.

The legendary romantic tale related with Athirapally Falls adds a heartfelt aspect to its social importance. The falls, as an observer to the timeless dance of adoration among water and shakes, become a representation for getting through connections. The nearby networks frequently integrate this subject into their social articulations, praising the never-ending connection among nature and humankind.

Dudhsagar Falls, named after the foamy appearance of its diving waters, holds own storehouse of stories reverberate with the social and authentic texture of Goa. The legend of the princess washing in the falls and emptying milk into its waters perseveres as an immortal story that loans the falls its suggestive name, Dudhsagar, or the "Ocean of Milk."

The Dudhsagar sanctuary, situated nearby, turns into a point of convergence for explorers who embrace an excursion to look for favors and submerge themselves in the consecrated waters. The falls, in this specific circumstance, are a display of nature as well as a course for otherworldly practices that have been imbued in the social customs of the locale for quite a long time.

Neighborhood legends likewise incorporates stories of experiences with extraordinary creatures close to Dudhsagar Falls. Accounts of strange lights and sounds radiating from the encompassing woods add to the supernatural atmosphere of the district. The falls, settled inside the Bhagwan Mahavir Untamed life Safe-haven, become a gathering point between the known and the obscure, where the normal and powerful domains blend.

The frontier history of Dudhsagar Falls, set apart by the now-old rail line tracks passing by its bluffs, adds a layer of verifiable importance to the regular miracle. The remainders of the rail route span, remaining as quiet observers to former times, summon a feeling of wistfulness and add to the falls' social legacy. The juxtaposition of human foundation with the crude excellence of nature turns into a sign of the consistently developing connection among individuals and their environmental elements.

Similarly as with Run Falls and Athirapally Falls, Dudhsagar Falls incorporates consistently into the social acts of the neighborhood networks. The love of divinities related with normal components, including water, is a typical social custom among the ancestral networks occupying the locale. The falls, as a sign of heavenly energy, become a holy space where customs and functions interface individuals with the profound substance of the scene.

These stories and neighborhood legends related with Run Falls, Athirapally Falls, and Dudhsagar Falls feature the significant manners by which these regular miracles are woven into the social personalities of their separate districts. They stop being simple geographical developments; all things being equal, they become archives of stories that mirror the profound connections among networks and the scenes they occupy.

The sharing of stories and nearby legends related with Run Falls, Athirapally Falls, and Dudhsagar Falls improves how we might interpret these regular marvels. These accounts rise above time, forming the shared awareness of individuals who live in vicinity to these fountains. As we drench ourselves in these stories, we find that these cascades are actual elements as well as living exemplifications of social legacy, profound convictions, and the persevering through association among people and the regular world.

Chapter 2

The Elegance of Western Ghats

The Western Ghats, a superb mountain range that traverses the western edge of the Indian subcontinent, arises as a demonstration of the perplexing dance among nature and topography. Extending roughly 1,600 kilometers from Gujarat to Kerala, the Ghats support an abundance of biodiversity, social variety, and remarkable scenes. This old mountain range, frequently alluded to as the "Sahyadri" in provincial dialects, holds a novel polish that unfurls across its undulating territories, thick timberlands, and dynamic environments.

The topographical history of the Western Ghats traverses a long period of time, set apart by structural movements, volcanic action, and the molding powers of disintegration. The starting points of the Ghats can be followed back to the separation of the supercontinent Gondwana, a land occasion that set up for the Indian subcontinent's excursion towards its current position. The structural plate developments brought about the inspiring of the Western Ghats, making a geological wonder that keeps on charming geologists, scientists, and nature devotees the same.

The Western Ghats is a gold mine of biodiversity, perceived as one of the world's eight "most sultry areas of interest" of natural variety. Its different biological systems, going from evergreen woodlands to fields, give a home to a stunning assortment of greenery.

The Ghats are home to an expected 7,402 types of blooming plants, 139 well evolved creature species, 508 bird species, and a large number of reptiles, creatures of land and water, and bugs. The locale's rich biodiversity isn't just a wellspring of logical interest yet additionally a social and biological legacy that has supported networks for ages.

The evergreen timberlands of the Western Ghats, portrayed by their rich shelters and different understory, contribute fundamentally to the district's environmental honesty. These timberlands go about as essential carbon sinks, assuming a critical part in managing the worldwide environment. The endemic species found in these woods, for example, the lion-followed macaque and the Malabar dim hornbill, add to the environmental uniqueness of the Ghats. The Shola meadows, scattered inside the montane backwoods, make a mosaic of natural surroundings that further upgrades the locale's biodiversity.

One of the unmistakable highlights of the Western Ghats is various streams and cascades overflow through the precarious slants. These water bodies, starting from the Ghats, not just extinguish the thirst of the fields underneath yet additionally shape the geology through disintegration and testimony. The Ghats are the well-spring of significant waterways like the Godavari, Krishna, Kaveri, and Tungabhadra, which stream toward the east, toward the west, and toward the south, depleting into the Sound of Bengal and the Bedouin Ocean. The organization of streams brings forth famous cascades, including the eminent Run Falls, shaping an essential piece of the locale's picturesque charm.

The social embroidery of the Western Ghats is woven with strings of assorted customs, dialects, and ways of life. The Ghats are possessed by different native networks, each with its one of a kind lifestyle well established in the regular environmental factors. The Agasthyamalai people group in the southern Ghats, for instance, follows age-old practices that include a profound association with the biodiversity of the district. The Todas of the Nilgiri Slopes and the Warlis of Maharashtra keep up with unmistakable social works on, mirroring the variety of human encounters molded by the Ghats.

Sanctuaries and hallowed forests settled in the Western Ghats give testimony regarding the profound importance credited to these mountains. The Agastyarkoodam top in the southern Ghats is considered hallowed by the Kani clans, and journeys to this pinnacle are an essential piece of their social and strict practices. The old sanctuary of Mahabaleshwar, settled in the Sahyadri range, is committed to Ruler Shiva and draws lovers looking for both otherworldly comfort and the all encompassing perspectives the area offers.

The Western Ghats likewise harbor archeological miracles that give looks into antiquated human advancements. The stone cut caverns of Ellora and Ajanta, both UNESCO World Legacy Destinations, grandstand the compositional brightness of old societies that flourished in the area.

These caverns, cut into the mountains, are embellished with complicated models and canvases that portray the strict and social accounts of their time. The Ghats, with their geographical developments, became a material for imaginative articulation as well as a safe-haven for profound examination.

The Ghats have additionally been observer to authentic occasions that formed the fates of domains and civic establishments. The essential area of the mountain range made it a characteristic obstruction, impacting shipping lanes and military procedures. The old shipping lane known as the Konkan Maurya course crossed the Ghats, interfacing the western coast with the Deccan level. The Ghats turned into a support of social trades, with brokers, pilgrims, and researchers moving between the waterfront locales and the inside.

The different environments and rises found in the Western Ghats add to the development of extraordinary biological systems, known as "sky islands." These are segregated peak environments that have species adjusted to explicit natural circumstances. As one climbs or dives the inclines of the Ghats, unmistakable microclimates arise, each supporting a specific local area of plants and creatures. These sky islands, with their endemic species and biological complexities, add to the district's status as a worldwide biodiversity area of interest.

The Ghats likewise assume a urgent part in directing the environment of the Indian subcontinent. The mountains block the dampness loaded breezes from the Bedouin Ocean, causing weighty precipitation on the windward side and making precipitation shadows on the leeward side. The occasional rainstorm, fundamental to the Indian environment, owe their examples to the presence of the Western Ghats. The Ghats impact the timing and power of the rainstorm, making them a climatic key part for the whole subcontinent.

Nonetheless, the Western Ghats are not immaculate by the difficulties of the cutting edge time. Human exercises, including deforestation, horticulture, and urbanization, present dangers to the delicate biological systems of the Ghats. The transformation of normal living spaces into farming grounds, particularly as tea and espresso ranches, has prompted environment fracture and loss of biodiversity. The unpredictable utilization of pesticides and manures in these estates further worsens the natural effect, influencing the sensitive equilibrium of the environments.

Mining exercises, especially for minerals and stones, have raised worries about the debasement of the Western Ghats. Quarrying activities can prompt soil disintegration, water contamination, and natural surroundings obliteration. The natural repercussions of such exercises reach out past the prompt mining destinations, influencing the whole scene and the biodiversity it maintains. Preservation endeavors and supportable land-use rehearses are vital to moderating these dangers and safeguarding the biological uprightness of the Ghats.

Environmental change represents an extra test toward the Western Ghats. The changes in temperature and precipitation examples can have flowing impacts on the area's biodiversity. Changes in the dispersion of plant and creature species, modified blooming and fruiting seasons, and changes in the planning of storms are a portion of the expected effects of environmental change on the Ghats. Preservation procedures need to consolidate environment versatility to guarantee the proceeded with endurance of the locale's interesting biological systems.

Regardless of these difficulties, the Western Ghats stand as a reference point of flexibility, biodiversity, and social lavishness. Protection drives, both administrative and local area drove, plan to alleviate the dangers and advance economical practices. Safeguarded regions, public parks, and untamed life asylums dab the scene, giving places of refuge to endemic species and considering the recovery of normal environments.

The tastefulness of the Western Ghats lies not just in its land highlights, biodiversity, and social legacy yet in addition in the interconnectedness of these components. The mountains, waterways, woods, and individuals structure a complicated trap of connections that shape the locale's personality. Protecting the style of the Western Ghats requires a comprehensive methodology that addresses biological, social, and financial viewpoints, perceiving the inherent worth of this regular wonder.

As we dive further into the complex polish of the Western Ghats, it becomes obvious that the district's appeal stretches out past its actual qualities. The mind boggling associations between the land, environmental, and social features of the Ghats unfurl a story that rises above logical request and embraces the embodiment of concurrence. The unique exchange among nature and mankind in this uneven territory is a demonstration of the sensitive harmony that supports life in the Western Ghats.

The Western Ghats act as a watershed for various waterways, supporting both the fields and the lower regions with their nurturing waters. The Bhadra Stream, beginning in the Ghats, is symbolic of this crucial job. It wanders through the scenes, cutting its direction through rich woods and rough landscapes, before in the end converging with the Tungabhadra Stream. The stream's process represents the interconnectedness of biological systems, delineating what the wellbeing of the Western Ghats means for the water security of immense areas downstream.

The biodiversity of the Western Ghats isn't restricted to the thick backwoods alone. The prairies, wetlands, and montane biological systems add to the district's environmental wealth. The Nilgiri tahr, an imperiled mountain goat, finds shelter in the verdant slants of the Nilgiri Slopes. The special verdure of the fields, adjusted to fire systems and occasional changes, adds to the biodiversity mosaic. The Ghats, with their elevational inclinations, harbor unmistakable vegetation zones, each supporting a particular local area of plant and creature species.

The woodlands of the Western Ghats likewise assume a significant part in moderating the effects of environmental change. The sequestration of carbon by the broad tree cover directs the worldwide environment. Moreover, the Ghats go about as a support against outrageous climate occasions, like floods and avalanches. The lavish coverings and root foundations of the trees forestall soil disintegration and add to the security of the scene. Perceiving the significance of these biological system administrations, protection endeavors mean to defend the regular strength of the Western Ghats notwithstanding an evolving environment.

The social variety implanted in the Western Ghats adds layers of extravagance to its tastefulness. The Ghats are home to different native networks, each with its remarkable customs, dialects, and information frameworks. The Kani clans of the Agasthyamalai locale, for example, have a close comprehension of the restorative properties of the plants that flourish in their environmental factors. Their customary environmental information, went down through ages, highlights the advantageous connection among people and the different greenery of the Ghats.

Holy forests, dissipated across the Western Ghats, are storehouses of social practices and biodiversity protection. These forests, safeguarded by neighborhood networks because of their strict and profound importance, act as microcosms of unblemished biological systems. The customs and functions led in these hallowed forests build up the possibility that nature isn't simply an asset to be taken advantage of yet a holy substance meriting love and security.

Sanctuaries settled in the Western Ghats mirror the mix of otherworldliness with the normal scene. The Marikamba Sanctuary in Sirsi, Karnataka, encompassed by thick woodlands, epitomizes this amicable relationship. Enthusiasts think about the regular components, like trees and shakes, as indications of the heavenly. The beneficial interaction of strict practices and natural mindfulness makes a novel social ethos that celebrates both otherworldly convictions and ecological stewardship.

The Western Ghats additionally take the stand concerning the union of customary and current rural practices. The terraced development on the inclines of the Ghats, known as "kaada panta" or wetland development, is a conventional strategy that uses the normal shapes of the land. These well established rural practices support neighborhood vocations as well as add to the protection of soil and water assets. In any case, the locale is additionally wrestling with current horticultural difficulties, including the development of monoculture ranches and the utilization of substance inputs.

The rich scenes of the Western Ghats have motivated workmanship, writing, and old stories for quite a long time. Artists and authors take care of drawn motivation from the Ghats' fog tops, thick woods, and flowing cascades. Crafted by celebrated creators like Kuvempu and Rabindranath Tagore reverberation the otherworldly and supernatural characteristics of the Ghats. The Western Ghats, with its immortal magnificence, fills in as a dream for imaginative articulations that catch the embodiment of the locale's style.

Nearby celebrations and festivities in the Western Ghats are profoundly laced with the normal rhythms of the scene. The Kodavas of Coorg praise the yearly Cauvery Sankramana celebration, denoting the waterway's rising up out of Talakaveri. The celebration is a sign of the personal connection between the local area and the waterway that supports their rural practices. Likewise, the Attukal Pongala celebration in Kerala is held in adoration to the goddess Attukal Bhagavathy, representing the holiness of the land and its assets.

Notwithstanding the natural and social meaning of the Western Ghats, the locale faces continuous dangers from human exercises. Impromptu urbanization, foundation advancement, and land-use changes present difficulties to the fragile equilibrium that has supported the Ghats for centuries. Preservation endeavors wrestle with the need to address these human-incited pressures while guaranteeing the prosperity of both the environments and the networks subject to them.

Biodiversity protection in the Western Ghats requires an exhaustive methodology that thinks about the necessities of the two individuals and nature. Local area based preservation drives engage nearby networks to become stewards of their regular legacy. By including native networks in protection endeavors, there is a potential chance to coordinate conventional information with current science, encouraging a comprehensive and reasonable way to deal with dealing with the Western Ghats' environments.

Safeguarded regions and natural life safe-havens, for example, Quiet Valley Public Park and Periyar Tiger Save, assume a crucial part in protecting the Western Ghats' biodiversity. These sanctuaries act as shelters for jeopardized species and consider logical exploration to unwind the secrets of the district's nature. The creation and viable administration of such safeguarded regions are fundamental parts of the protection system for the Ghats.

Instructive drives and mindfulness crusades additionally add to the preservation ethos of the Western Ghats. By ingraining a feeling of satisfaction and obligation among nearby networks, progressives intend to encourage an aggregate obligation to defending the locale's regular fortunes. Natural schooling programs draw in networks in grasping the biological complexities of the Ghats and advance economical practices that blend with the scene.

Economical the travel industry, when overseen capably, can be an instrument for both protection and local area improvement in the Western Ghats. By giving financial open doors to neighborhood networks, the travel industry drives can boost the security of regular assets. Nonetheless, cautious preparation and adherence to supportable practices are essential to forestall the adverse consequences of the travel industry on delicate biological systems.

Logical exploration in the Western Ghats keeps on disclosing new disclosures and develop how we might interpret its natural intricacies.

From recording new species to concentrating on environmental change influences, specialists contribute important experiences that illuminate protection techniques. The joining of logical discoveries with conventional information makes a vigorous starting point for dealing with the Ghats' biodiversity in a quickly impacting world.

The class of the Western Ghats resounds not just in its geographical arrangements, biodiversity, and biological administrations yet additionally in the complicated connections woven into its scenes. The social variety, conventional practices, and otherworldly associations highlight the profound reliance among people and nature. As the district explores the difficulties of the cutting edge time, an agreeable harmony between preservation, local area prosperity, and feasible improvement is crucial for save the persevering through tastefulness of the Western Ghats for a long time into the future.

2.1 Focus on the Western Ghats, a hotspot for waterfalls in India.

The Western Ghats, a stunning mountain range along the western shore of India, remains as an entrancing material of biodiversity and land ponders. Among its many captivating highlights, the Western Ghats arise as a focal point for cascades, exhibiting nature's loftiness and the unpredictable dance among mountains and water. As the rainstorm twists clear across the Ghats, they deliver an outpouring of cascades that decorate the scene, making an orchestra of sights and sounds that enthrall the faculties.

The variety of cascades in the Western Ghats is a demonstration of the locale's geological complexities. From the northern compasses of Gujarat toward the southern tip of Kerala, the Ghats have a bunch of cascades, each with its interesting person and appeal. As the storm downpours soak the thick woodlands and rough bluffs, the Western Ghats change into a sanctuary of flowing waters, leading to a portion of India's most notorious and stunning cascades.

Run Falls, settled in the core of the Western Ghats in Karnataka, remains as a great model of the locale's cascade grandness. Made by the Sharavathi Waterway, Run Falls is a demonstration of the crude force of nature. Plunging from a level of around 830 feet, it is one of the greatest untiered cascades in India. The moniker "Run" is gotten from the Kannada word for "thunder," a fitting descriptor for the deafening sound made by the falling waters. The four unmistakable fountains — Raja, Rani, Roarer, and Rocket — add to the falls' sensational charm, making a scene that leaves a permanent engraving on the spectator's memory.

Past its sheer level and volume, Run Falls additionally catches the social and otherworldly creative mind of the neighborhood people. The falls are profoundly laced with the legends of the locale, filling in as a dream for writers and craftsmen the same. Legends trait divine starting points to Run Falls, taking into account the Sharavathi Waterway as the natural appearance of the goddess Sharavathi herself.

Pioneers frequently view the falls as a consecrated site, representing the goddess' plunge to favor the land. The social reverberation of Run Falls stretches out to neighborhood celebrations, where the flowing waters become a vital piece of customs and festivities.

Further south in Kerala, Athirapally Falls unfurls as a verdant work of art in the midst of the rich scenes of the Western Ghats. Frequently alluded to as the "Niagara of India," Athirapally Falls is a stunning sight where the Chalakudy Stream slips from the Anamudi mountains. The falls, with their sweeping width and voluminous stream, make a visual display that has made them an inclined toward background for Indian film.

Athirapally Falls isn't only a geological wonder; it is likewise cherished in neighborhood legends and ancestral old stories. The stream spirits, charmed by the falls' excellence, are accepted to accumulate here to observe the timeless dance of adoration among water and shakes. The falls are viewed as a gathering point for divine creatures, adding a mysterious aspect to their charm. The Kadar clan, native to the district, sees Athirapally Falls as a hallowed space, meshing their own stories into the social texture of the Ghats. The realistic noticeable quality of Athirapally Falls has additionally raised its status, drawing vacationers and producers the same to observe the amicable marriage of nature and workmanship.

Moving towards the southwestern tip of the Western Ghats, the Dudhsagar Falls in Goa catches the creative mind with its foamy, flowing waters. Meaning the "Ocean of Milk," Dudhsagar Falls is a scene that unfurls in the midst of the Bhagwan Mahavir Untamed life Safe-haven. The falls, plunging from a level of around 1,017 feet, make a dreamlike vista as the waters change into a foamy, white mass during their plummet.

Neighborhood legends encompass Dudhsagar Falls, adding a dash of persona to its charming excellence. One such story portrays the tale of a princess who used to wash in the falls, pouring a container of sweet milk into the waters prior to stepping in. The milk, converging with the falling water, made the deception of an ocean of milk, giving the falls their suggestive name. This legendary account loans Dudhsagar Falls its social importance as well as adds to the mysterious emanation that encompasses the area.

The topographical developments of the Western Ghats assume a significant part in forming these fantastic cascades. The undulating territory, described by steep slopes and rough outcrops, gives the ideal material to the flowing waters. The waterways and streams that begin in the high-height districts of the Ghats explore through thick woods and rough scenes, picking up speed prior to making their sensational plummet into the valleys beneath.

The biodiversity of the Western Ghats further improves the appeal of its cascades. The thick backwoods that shroud the mountains add to the unblemished nature of the water, making a living space for different greenery.

The encompassing biological systems, including endemic plant species and uncommon natural life, add an environmental aspect to the cascades, making them vital parts of the Ghats' rich biodiversity.

The environmental meaning of these cascades reaches out past their nearby environmental elements. The wet conditions made by the fog and shower from the falling waters add to the arrangement of microclimates, encouraging one of a kind plant and creature networks. Greeneries, plants, and orchids track down a specialty in the clammy hole around the falls, making a fragile biological system that blossoms with the never-ending dampness produced by the flowing waters.

The Western Ghats, assigned as an UNESCO World Legacy Site, is perceived for its remarkable natural variety and environmental significance. The cascades, as powerful parts of this environmental embroidery, assume a part in supplement cycling and keeping up with the natural equilibrium of the district. They add to the hydrological cycle, affecting atmospheric conditions and precipitation conveyance across the subcontinent.

The occasional varieties in the Western Ghats, set apart by the storm downpours, enhance the scene of its cascades. The rainstorm, which regularly happens from June to September, changes the scene into a rich green material. The downpour took care of waterways and streams grow, arriving at their pinnacle stream, and the cascades become completely awake with reestablished power. The rainstorm season is a period of restoration for the Western Ghats, renewing its water assets and supporting the different environments that flourish in its hug.

While the rainstorm delivers the glory of the cascades, it additionally presents difficulties to the locale. The precarious inclines and weighty precipitation make the Ghats vulnerable to avalanches and soil disintegration. Protection endeavors in the Western Ghats need to find some kind of harmony between tackling the biological advantages of the rainstorm and moderating the dangers related with outrageous climate occasions.

The Western Ghats are likewise a pivotal wellspring of water for the fields and lower regions encompassing the mountain range. The streams beginning in the Ghats add to the water supply of various urban communities, horticultural terrains, and biological systems downstream. The cascades, with their perpetual stream, add to the nonstop recharging of these waterways, framing a help for both human networks and the normal world.

As the Western Ghats face different anthropogenic tensions, including deforestation, urbanization, and horticultural extension, the protection of its cascades becomes basic. Reasonable land-use rehearses, afforestation drives, and the conservation of normal environments are significant parts of guaranteeing the drawn out strength of the Western Ghats and its notable cascades.

Preservation endeavors ought to likewise consolidate local area commitment and mindfulness projects to encourage a feeling of obligation and stewardship among the nearby people.

The travel industry, driven by the appeal of the Western Ghats' cascades, presents the two open doors and difficulties. While dependable the travel industry can add to the neighborhood economy and make mindfulness about the locale's natural significance, unrestrained the travel industry can prompt ecological corruption and territory aggravation. The execution of maintainable the travel industry works on, including controlled appearance, squander the board, and the advancement of eco-accommodating other options, is crucial for figure out some kind of harmony among preservation and the travel industry.

Digging further into the charm of the Western Ghats' cascades uncovers a story of biological complexities, social importance, and the difficulties looked by these normal marvels in the cutting edge time. Past their visual magnificence, these flowing waters are strings woven into the texture of the Ghats, molding the locale's environment, rousing social stories, and featuring the fragile harmony among nature and human communication.

The Western Ghats, assigned as a biodiversity area of interest, harbor a rich embroidery of verdure, quite a bit of which is complicatedly connected to the cascades that intersperse the scene. The soggy microclimates made by the fog and shower around cascades cultivate novel biological systems, giving a shelter to greeneries, plants, and orchids. The lasting progression of water adds to the food of oceanic life, including fish and spineless creatures, making complicated food networks that echo through the Ghats' environments.

The cascades likewise assume an essential part in the hydrological pattern of the district. The flowing waters re-energize groundwater supplies, adding to the enduring progression of waterways that start in the Ghats. This hydrological importance stretches out to the fields and lower regions encompassing the Western Ghats, where streams go about as helps for agribusiness, human settlements, and various biological systems downstream. The occasional varieties in cascade stream, directed by the rainstorm, are basic to keeping up with this complex equilibrium in the water cycle.

Run Falls, with its transcending presence, epitomizes the geographical wonders that add to the development of these cascades. The Sharavathi Waterway, slipping from the Deccan Level, experiences the sudden ledge of the Western Ghats, making the ideal setting for the sensational dive of Run Falls. The rough territory and steep bluffs channel the power of the stream into a stunning display that exhibits the topographical powers forming the scene.

The geography of the Western Ghats, portrayed by old rocks, separation points, and volcanic leftovers, assumes a crucial part in deciding the structure and area of its cascades. The layered basalt developments give a material to the flowing waters to draw their direction through, making unpredictable examples and creeks that characterize the personality of every cascade.

The powerful collaboration between rock developments and water stream adds to the chiseling of the Ghats' scenes over centuries.

The social meaning of cascades in the Western Ghats rises above their environmental job, meshing into the customs and stories of the nearby networks. These cascades are not simply topographical elements; they are worshipped as signs of the heavenly, images of richness, and wellsprings of otherworldly motivation. The stories and fantasies encompassing Run Falls, Athirapally Falls, and Dudhsagar Falls are implanted in the social texture of the districts they beauty.

In Karnataka, Run Falls isn't just a topographical wonder yet additionally an otherworldly safe house. The Sharavathi Waterway, considered hallowed, turns into a channel for the heavenly at Run Falls. The falls are unpredictably connected to neighborhood celebrations and customs, where travelers leave on excursions to look for the favors of the goddess Sharavathi. The social festivals that unfurl against the scenery of Run Falls intensify its importance, making it a point of convergence for mutual character and otherworldly worship.

In Kerala, Athirapally Falls resounds with social articulations established in the customs of the native networks. The ancestral old stories, went down through ages, interweaves with the biological character of the falls. The watchman spirits, Kumara Yakshi and Kumara Yakshi, become piece of the social story, mirroring an agreeable connection between the otherworldly and normal domains. Conventional dance structures, like Theyyam and Kathakali, integrate the falls into their exhibitions, consolidating craftsmanship and nature in a consistent articulation of social extravagance.

Dudhsagar Falls in Goa is wrapped in a legendary romantic tale that adds a heartfelt aspect to its social importance. The princess washing in the falls and emptying milk into its waters perseveres as an immortal story, loaning the falls their suggestive name. The Dudhsagar sanctuary close by turns into a position of journey, where enthusiasts look for endowments and submerge themselves in the consecrated waters. The falls, encompassed by the Bhagwan Mahavir Untamed life Safe-haven, become a gathering point between the known and the obscure, where regular and powerful domains unite.

As the travel industry picks up speed in the Western Ghats, driven by the charm of its cascades, it brings the two open doors and difficulties. The inundation of guests looking to observe the greatness of Run Falls, experience the realistic magnificence of Athirapally Falls, or wonder about the foamy outpouring of Dudhsagar Falls has monetary ramifications for nearby networks. The travel industry can give vocations, encourage mindfulness about the biological significance of the locale, and add to protection endeavors.

Notwithstanding, the flood in the travel industry likewise raises worries about the natural effect on these delicate biological systems. Unregulated appearance can prompt living space aggravation, contamination, and the debasement of the normal scenes around cascades. Protection drives should explore the sensitive harmony between advancing the travel industry as a financial driver and guaranteeing the biological maintainability of these regular marvels.

Feasible the travel industry rehearses become basic to address the difficulties presented by expanding guest numbers. Controlled section, squander the board, and instructive projects can assist with limiting the biological impression of the travel industry. Neighborhood people group should be dynamic members in the travel industry story, receiving the rewards while effectively adding to the safeguarding of their regular legacy.

Protection endeavors in the Western Ghats should stretch out past individual cascades to envelop whole biological systems. Safeguarded regions, public parks, and untamed life safe-havens inside the Ghats act as basic shelters for biodiversity. Quiet Valley Public Park, Periyar Tiger Hold, and Bhagwan Mahavir Natural life Safe-haven are only a couple of instances of the preservation asylums that defend the different greenery of the district.

Logical examination assumes a vital part in grasping the natural complexities of the Western Ghats and figuring out successful preservation methodologies. Concentrates on biodiversity, hydrology, and environment flexibility contribute significant bits of knowledge that guide policymakers and progressives. The combination of conventional information, held by native networks, with logical discoveries makes an all encompassing comprehension of the perplexing elements at play in the Ghats.

Environmental change represents an extra layer of intricacy to the protection challenges looked by the Western Ghats. Modifications in temperature, precipitation designs, and the recurrence of outrageous climate occasions can affect the locale's biodiversity and water assets. Protection techniques need to consolidate environment flexibility, guaranteeing that the biological systems of the Western Ghats can adjust to changing climatic circumstances.

2.2 Explore the lush greenery, biodiversity, and the symbiotic relationship between the waterfalls and the ecosystem.

Leaving on an excursion through the Western Ghats divulges a scene hung in lavish vegetation, where biodiversity flourishes as one with flowing cascades. This locale, eminent for its natural lavishness, arises as a safe-haven of interconnected biological systems, where the cooperative connection among cascades and the climate lays out a representation of strength and conjunction.

The Western Ghats, an UNESCO World Legacy Site, stretches out along the western edge of the Indian subcontinent, enveloping states like Gujarat, Maharashtra, Goa, Karnataka, Kerala, and Tamil Nadu. Its undulating landscape, sprinkled with thick woodlands, fields, and wandering waterways, gives an optimal setting to the assembly of biodiversity and cascades. As the storm mists assemble, the Ghats wake up, changing into a verdant sanctuary that reverberations with the ensemble of streaming waters.

One of the meaningful cascades that accentuate the verdure of the Western Ghats is Run Falls in Karnataka. Settled in the midst of the thick backwoods, Run Falls remains as a demonstration of the Ghats' topographical ability and natural variety. The Sharavathi Waterway, beginning from the core of the Ghats, tears down the rough precipices, making an exhibition of unrivaled greatness. The sheer power of the falling waters, expanded by the storm downpours, cuts through the rough scene, molding the course of the waterway and adding to the district's hydrological balance.

The lavish vegetation encompassing Run Falls isn't simply a background however a fundamental piece of the cascade's charm. The Western Ghats, with its evergreen and deciduous timberlands, turns into a material for a rich embroidery of greenery that reaches from transcending trees to sensitive plants. The thick overhang not just adds to the visual scene of Run Falls yet additionally fills in as a pivotal part of the cascade's biological system. The vegetation assumes an essential part in managing the progression of water, forestalling soil disintegration, and keeping up with the soundness of the encompassing scene.

The harmonious connection among cascades and the biological system is additionally exemplified by the different vegetation that flourishes in the wet conditions made by the sliding waters. Greeneries, plants, and orchids track down asylum in the fissure of rocks, shaping sensitive environments supported by the unending fog and shower produced by Run Falls. These microhabitats add to the general biodiversity of the Ghats, making specialties for species adjusted to the interesting circumstances around cascades.

As we venture toward the south through the Western Ghats, Athirapally Falls in Kerala arises as one more gem in this biological crown. Frequently alluded to as the "Niagara of India," Athirapally Falls is an enamoring scene where the Chalakudy Stream plunges over steep precipices, making a display that catches the pith of the Ghats' normal magnificence. The falls, settled inside the hug of the Sholayar Reach, become a union point for biodiversity, water, and culture.

The thick vegetation encompassing Athirapally Falls envelops one of the final plots of tropical evergreen rainforest in the Western Ghats. This biological system, portrayed by its high species variety and endemic greenery, shapes a basic piece of the falls' environmental embroidery. The different exhibit of plant species, from transcending trees to understory bushes, adds to the exceptional microclimates that characterize the district around Athirapally Falls.

The biodiversity of the Western Ghats stretches out to the oceanic biological systems supported by its cascades. The Chalakudy Stream, revived by the diving waters of Athirapally Falls, turns into a help for heap fish species and spineless creatures. These freshwater biological systems, unpredictably associated with the Ghats' scene, add to the locale's generally speaking environmental flexibility. Preservation endeavors around Athirapally Falls perceive the significance of safeguarding the actual cascade as well as the more extensive biological systems that rely upon its perpetual stream.

Dudhsagar Falls, arranged in the biodiversity focal point of Goa, adds an alternate aspect to the harmonious connection among cascades and the biological system. Meaning the "Ocean of Milk," Dudhsagar Falls plunges emphatically from a level of around 1,017 feet, making a display that dazzles the creative mind. The falls, encompassed by the rich environs of the Bhagwan Mahavir Untamed life Safe-haven, become a microcosm of natural interconnectedness.

The tropical timberlands encompassing Dudhsagar Falls act as basic territories for different verdure. The falls' perpetual stream, supported by the storm down-pours, adds to the development of different microhabitats that help many species. The Western Ghats, eminent for its elevated degrees of endemism, sees this peculiarity reflected in the biological systems around Dudhsagar Falls. The safe-haven turns into a shelter for animal types adjusted to the mosaic of living spaces made by the falls, from the oceanic conditions to the encompassing earthbound scenes.

The interconnectedness of cascades and environments in the Western Ghats reaches out past the domains of vegetation to the mind boggling hydrological cycle that characterizes the district. The thick vegetation, normal for the Ghats, assumes a critical part in impacting the progression of water, forestalling soil disintegration, and keeping up with the steadiness of the scene. The cascades, thusly, add to the renewal of groundwater supplies, guaranteeing a constant progression of water to the streams that confuse the Western Ghats.

The Bhadra Stream, starting in the Western Ghats, represents the hydrological meaning of these cascades. Wandering through the scenes of Karnataka, the stream turns into a demonstration of the multifaceted dance among mountains and water. The flowing waters of Run Falls, which feed into the Bhadra Waterway, make a dynamic hydrological framework that shapes the biology of the district. The stream turns into a life saver, supporting rural grounds, human settlements, and different biological systems downstream.

The Western Ghats, frequently alluded to as the "Water Pinnacle of South India," impacts the territorial environment and atmospheric conditions through its hydrological commitments. The cascades, going about as regular controllers, help in the conveyance of precipitation, forestalling limits and guaranteeing a consistent stockpile of water to the fields and lower regions encompassing the Ghats. The occasional varieties in cascade stream, directed by the storm, become essential to the biological rhythms of the Western Ghats.

The social meaning of cascades in the Western Ghats reflects the personal connection between nearby networks and the environments they possess. The Kani clans in the Agasthyamalai district, for example, have a significant comprehension of the restorative properties of plants that flourish nearby cascades. Their conventional biological information, went down through ages, highlights the advantageous association among people and the different greenery of the Ghats.

Sacrosanct forests, dispersed across the Western Ghats, become archives of social practices and biodiversity protection. These forests, safeguarded by nearby networks because of their strict and otherworldly importance, act as microcosms of flawless biological systems. The customs and functions led in these hallowed forests build up the possibility that nature isn't simply an asset to be taken advantage of however a consecrated element meriting love and security.

Sanctuaries settled in the Western Ghats mirror the combination of otherworldliness with the normal scene. The Marikamba Sanctuary in Sirsi, Karnataka, encompassed by thick woods, epitomizes this agreeable relationship. Lovers think about the regular components, like trees and shakes, as appearances of the heavenly. The beneficial interaction of strict practices and natural mindfulness makes an interesting social ethos that celebrates both otherworldly convictions and ecological stewardship.

Nearby celebrations and festivities in the Western Ghats are profoundly laced with the regular rhythms of the scene. The Kodavas of Coorg commend the yearly Cauvery Sankramana celebration, denoting the waterway's rising up out of Talakaveri. The celebration is an indication of the close connection between the local area and the waterway that supports their farming practices. Also, the Attukal Pongala celebration in Kerala is held in worship to the goddess Attukal Bhagavathy, representing the holiness of the land and its assets.

The tastefulness of the Western Ghats lies not just in its land highlights, biodiversity, and social legacy yet additionally in the interconnectedness of these components. The mountains, streams, backwoods, and individuals structure a perplexing trap of connections that shape the locale's personality. Protecting the style of the Western Ghats requires a comprehensive methodology that addresses environmental, social, and financial viewpoints, perceiving the inborn worth of this normal wonder.

As we dig further into the many-sided connection among cascades and the environment of the Western Ghats, the story unfurls to uncover the difficulties and open doors that characterize the preservation scene of this exceptional district. In spite of its environmental extravagance, the Western Ghats faces continuous dangers from human exercises, requiring a coordinated work to figure out some kind of harmony among improvement and the safeguarding of its normal legacy.

Impromptu urbanization, foundation improvement, and agrarian development present critical difficulties to the fragile harmony of the Western Ghats. The developing human impression applies strain on the district's biological systems, prompting deforestation, living space fracture, and loss of biodiversity. As human settlements infringe upon the regular territories encompassing cascades, the delicate biological systems that support these flowing waters face the gamble of corruption.

Protection endeavors in the Western Ghats wrestle with the need to address these human-actuated pressures while guaranteeing the prosperity of both the environments and the networks subject to them. Perceiving the many-sided linkages among nature and human livelihoods, preservation drives progressively embrace an all encompassing methodology that coordinates the requirements of neighborhood networks with the objectives of biodiversity protection.

Local area based protection drives engage nearby populaces to become stewards of their regular legacy. Drawing in native networks, like the Kani clans in the Agasthyamalai locale, in preservation endeavors jelly customary biological information as well as encourages a feeling of pride and obligation. The cooperative connection among cascades and nearby societies is consequently woven into the texture of protection works on, guaranteeing that these normal miracles persevere as a wellspring of pride and food for a long time into the future.

Safeguarded regions and untamed life safe-havens assume a significant part in protecting the Western Ghats' biodiversity. Quiet Valley Public Park, Periyar Tiger Save, and Bhagwan Mahavir Untamed life Safe-haven act as shelters for jeopardized species and focal points for logical exploration. These safeguarded regions become shelters where the interconnectedness of cascades, biological systems, and natural life is protected, offering a brief look into the sensitive equilibrium that characterizes the Western Ghats.

Instructive drives and mindfulness crusades add to the protection ethos of the Western Ghats. By imparting a deep satisfaction and obligation among nearby networks, traditionalists expect to cultivate an aggregate obligation to protecting the district's normal fortunes. Natural schooling programs draw in networks in grasping the biological complexities of the Ghats and advance maintainable practices that fit with the scene.

Feasible the travel industry, when overseen mindfully, arises as an instrument for both protection and local area improvement in the Western Ghats. By giving financial open doors to nearby networks, the travel industry drives can boost the insurance of regular assets. Nonetheless, cautious preparation and adherence to supportable practices are vital to forestall the adverse consequences of the travel industry on delicate environments. The harmony between advancing the travel industry and guaranteeing the safeguarding of the Western Ghats' natural respectability requires insightful and vital administration.

Logical examination in the Western Ghats keeps on disclosing new disclosures and develop how we might interpret its natural intricacies. From reporting new species to concentrating on environmental change influences, scientists contribute important bits of knowledge that illuminate protection methodologies. The mix of logical discoveries with customary information makes a vigorous starting point for dealing with the Ghats' biodiversity in a quickly impacting world.

In confronting the difficulties presented by environmental change, the Western Ghats should adjust to modified temperature and precipitation designs. Preservation procedures need to integrate environment strength, guaranteeing that the biological systems of the Ghats can endure and recuperate from the effects of an evolving environment. The cascades, as powerful parts of the locale's hydrological cycle, become marks of the more extensive biological flexibility expected for the Western Ghats to persevere even with worldwide natural changes.

2.3 Highlight lesser-known gems hidden within the Western Ghats.

Setting out on an excursion into the core of the Western Ghats discloses a mother lode of less popular diamonds, concealed in the midst of the verdant scenes and transcending tops. While Run Falls, Athirapally Falls, and Dudhsagar Falls stand as notorious images of the district's greatness, there exist heap less popular cascades that weave a nuanced story of the Ghats' secret magnificence. These disguised fountains, frequently concealed in remote corners, offer a brief look into the immaculate and unseen features of this UNESCO World Legacy Site.

In the northern scopes of the Western Ghats, settled inside the thick timberlands of Gujarat, lies the spellbinding Gira Falls. Gira Falls, otherwise called the Waghai Cascades, is a less popular pearl that exudes a quiet appeal. The Purna Waterway, twisting through the Darn region, dives from the Sahyadri Reach, making this beautiful fountain. Encircled by teak woods and undulating scenes, Gira Falls is a shelter of quietness, immaculate by the clamoring swarms that frequently run to additional renowned partners.

What recognizes Gira Falls is its openness, offering a quiet retreat for those looking for isolation in the midst of nature. The falls, however unobtrusive in level contrasted with a portion of its partners, ooze an unpretentious style as the waters overflow over the rough edges. The encompassing vegetation, overflowing with biodiversity, makes a verdant scenery that supplements the delicate progression of the falls. Gira Falls turns into a safe-haven where the orchestra of streaming waters fits with the stirring leaves and the calls of inconspicuous untamed life.

Wandering further into the Ghats, one experiences the unlikely treasure of Kune Falls in Maharashtra. Concealed in the Lonavala-Khandala valley, Kune Falls stays a moderately lesser-investigated cascade in spite of its spellbinding excellence.

The perpetual stream beginning from the Amba Waterway plunges from a level of roughly 200 feet, making an entrancing exhibition as it slips through the rich vegetation.

What separates Kune Falls is its three-layered structure, with each outpouring adding to the falls' novel appeal. The highest level, the most elevated of the three, charms with its sheer drop, while the resulting levels make fragile cover of water that add to the by and large visual allure. The encompassing scene, enhanced with a blend of native verdure, intensifies the charm of Kune Falls, making it a secret desert garden for nature fans.

In the southern domains of the Western Ghats, the less popular Sathodi Falls in Karnataka coaxes with its untainted excellence. Situated in the Uttara Kannada area, Sathodi Falls stays confined, settled inside the thick evergreen backwoods that describe this district. The cascade is framed by the conjunction of a few anonymous streams, making a fountain that plunges over rough outcrops, shaping a characteristic pool at its base.

Sathodi Falls embodies the unlikely treasures inside the Western Ghats, open just through winding woods trails that lead to its immaculate environs. The disconnection of this cascade adds to its unsullied appeal, offering guests a valuable chance to drench themselves in the immaculate excellence of the Ghats. The perfectly clear pool shaped by the falling waters turns into a serene desert spring, welcoming the people who adventure off in an unexpected direction to relish an experience of isolation in nature's hug.

The Western Ghats harbor flowing cascades as well as less popular regular miracles, for example, the secretive Yana Rocks in Karnataka. While not a customary cascade, Yana Rocks are a fundamental piece of the Ghats' secret fortunes. Settled in the midst of the thick backwoods of the Sahyadri Reach, these novel limestone rock developments rise decisively from the earth, making a strange scene that feels practically powerful.

The Bhairaveshwara Shikhara and the Mohini Shikhara, the two conspicuous stone developments at Yana, give testimony regarding the powers of enduring that have etched them into their ongoing shapes. Encircled by thick vegetation, these stones radiate a quality of persona and act as a consecrated site for neighborhood networks. The secret caverns and sinkholes in the area add to the charm of Yana, making it an objective that goes past the regular cascade insight.

Moving towards the Kerala-Tamil Nadu line, the less popular Perunthenaruvi Falls unfurls as an unlikely treasure in the Western Ghats. This outpouring, arranged on the Pamba Stream, is portrayed by its thin and prolonged structure as the water dives smoothly over a rough incline. Perunthenaruvi, meaning "The Incomparable Honey Stream," implies the brilliant tint of the cascade during the storm season, making a visual display that adds to its charm.

Perunthenaruvi Falls, however not quite as generally celebrated as a portion of its partners, offers an exceptional viewpoint on the different structures that cascades can take inside the Western Ghats. The encompassing scenes, decorated with emerald vegetation, add to the falls' beautiful setting. The general peacefulness of this unexpected, yet invaluable treasure gives a break to those looking for a calmer and more close fellowship with nature.

In the Coimbatore region of Tamil Nadu, the less popular Kovai Kutralam Falls adds a hint of serenity toward the Western Ghats' collection of stowed away ponders. Settled inside the Siruvani Slopes, this cascade is taken care of by the Siruvani Waterway, and its immaculate magnificence remains moderately neglected by standard the travel industry. The falls overflow tenderly over the rough landscape, making a grand scene that mirrors the quietness of its environmental factors.

Kovai Kutralam Falls, encompassed by rich timberlands, is important for the Siruvani Dam's catchment region, guaranteeing the immaculateness of its waters. The falls become a secret safe-haven for those looking for a tranquil retreat, away from the groups that frequently describe more famous objections. The untrodden ways prompting Kovai Kutralam offer a door to isolation and normal wonder.

As we cross the length and broadness of the Western Ghats, experiencing these less popular jewels, it becomes clear that the district isn't simply an assortment of renowned cascades however an immense material of stowed away ponders ready to be found. The appeal of these covered fountains lies in their visual allure as well as in the accounts they tell — accounts of perfect scenes, immaculate by the clamoring scene past.

These unlikely treasures highlight the requirement for dependable and supportable the travel industry works on, guaranteeing that the sensitive environments encompassing these cascades stay safe. Preservation endeavors, established in a profound comprehension of the interconnectedness of verdure, fauna, and water, are crucial for protect the Western Ghats' secret fortunes for people in the future.

Wandering further into the hid scenes of the Western Ghats, one experiences the entrancing appeal of Thoseghar Falls in Maharashtra. Concealed in the Satara region, Thoseghar Falls stays a secret gem in the midst of the Sahyadri Reach. The falls overflow down a progression of steps, making an all encompassing scene that reverberates with the quietness of its environmental elements. While Thoseghar isn't quite so famous as a portion of its partners, its downplayed excellence and the encompassing lavish plant life make it a dazzling objective for those looking for isolation in nature's hug.

The outpouring of Thoseghar Falls acquires unmistakable quality during the storm season when the downpours revive the scene, changing the environmental elements into a lively embroidery of green.

The falls, encompassed by thick woods and delegated by cloudy cloak, become a sanctuary for biodiversity. The different greenery that flourishes in the area adds to the falls' charm, making a living space for various species that track down shelter in this unblemished climate.

One more unlikely treasure inside the Western Ghats uncovers itself as Meenmutty Falls in Kerala. Concealed in the Wayanad locale, Meenmutty Falls is a demonstration of the district's immaculate magnificence. The cascade, took care of by the Meenmutty Stream, drops from a level of around 300 meters, making a three-layered overflow that dives into a pool beneath. What separates Meenmutty Falls is the excursion one should embrace to arrive at its perfect environs — a trip across the lavish woodlands that adds a bold aspect to the experience.

Meenmutty Falls encapsulates the possibility of a secret desert spring, where the ensemble of falling waters blends with the stirring leaves and the calls of endemic birds. The journey to the falls permits explorers to cross through scenes embellished with assorted vegetation, making a vivid encounter that goes past the visual wonder of the actual cascade. Meenmutty Falls, with its isolated appeal, turns into a retreat for the individuals who look for the excitement of investigation combined with the quietness of nature.

A less popular wonder settled in the core of Karnataka's Western Ghats is the unlikely treasure of Hebbe Falls. Arranged close to Kemmangundi, Hebbe Falls stays hidden by the woods of the Bhadra Untamed life Asylum, adding to its emanation of segregation. The falls are separated into two phases — the Dodda Hebbe (Enormous Falls) and Chikka Hebbe (Little Falls) — each offering its own charming viewpoint on the flowing waters.

What makes Hebbe Falls vital is the excursion one embraces to arrive at this secret heaven. Open just through a jeep ride followed by a trip across the thick woods, the falls hold a demeanor of selectiveness. The encompassing scenes, embellished with a different scope of plant species, make a characteristic amphitheater that improves the falls' visual allure. Hebbe Falls, with its distant area and immaculate setting, stays a very much protected secret inside the Western Ghats.

Further south, in the Coorg locale of Karnataka, the quiet Chelavara Falls arises as a less popular fortune. Taken care of by the Virajpet Thadiyandamol Slope Reach, Chelavara Falls flows effortlessly over rough outcrops, making a quiet environment that differentiations with the deafening presence of additional popular cascades. The falls are encircled by espresso estates and rich scenes, adding to the general appeal of the objective.

Chelavara Falls, however not as fabulous in that frame of mind, with its effortlessness and the untainted excellence of its environmental elements. The rough landscape around the falls makes normal pools, welcoming guests to submerge themselves in the cool waters while absorbing the quiet feeling.

The secret charm of Chelavara Falls lies in its capacity to offer a rest from the clamor of famous vacationer locations, permitting voyagers to associate with the pure embodiment of the Western Ghats.

As we navigate the Ghats, the less popular diamonds stretch out past cascades to uncover the charming environment of Shola Woods. Shola Woods, an extraordinary montane environment tracked down in the higher elevations of the Western Ghats, remain somewhat neglected contrasted with their popular partners. These cloud backwoods, described by hindered trees, dynamic greenery covered rocks, and a rug of endemic vegetation, add to the environmental variety that characterizes the Ghats.

The secret magnificence of Shola Woodlands lies in their job as supplies of biodiversity. These backwoods harbor various endemic species, including uncommon orchids, greeneries, and restorative plants. The harmonious connection between Shola Timberlands and contiguous prairies makes a unique scene where biodiversity flourishes. The Ghats' less popular miracles, like Shola Backwoods, underline the requirement for thorough preservation endeavors that envelop cascades as well as the complex biological systems that support them.

Traveling towards the southern tip of the Western Ghats, the less popular Agasthyamalai Biosphere Hold unfurls as a preservation safe house. Riding the boundary of Kerala and Tamil Nadu, this biosphere save remains somewhat dark contrasted with other safeguarded regions in the area. The save envelops the Agasthyamalai Reach, which is loved as the home of Agastya Muni, a worshipped sage in Hindu folklore.

The Agasthyamalai Biosphere Save turns into a safe-haven for different vegetation, including a few endemic and imperiled species. The scenes inside the save range from fields and shola woodlands to tropical evergreen backwoods, making a mosaic of natural surroundings. The presence of uncommon and endemic species, for example, the Nilgiri tahr and the lion-followed macaque, highlights the significance of saving this less popular diamond inside the Western Ghats.

The story of the Western Ghats' secret miracles stretches out past cascades and safeguarded regions to the rich woven artwork of native societies. The less popular Malnad locale in Karnataka, settled inside the Western Ghats, is a social mother lode that remains clouded by the shadows of its more well known partners. Malnad, meaning "place where there is slopes," is portrayed by its novel practices, dynamic celebrations, and culinary enjoyments that mirror the nearby lifestyle.

The confined appeal of Malnad lies in its flourishing rural practices, where espresso and flavors prosper in the Ghats' prolific soil. The district turns into a safe house for those looking for a social inundation, where customary homes, known as "ainmans," stand as demonstrations of the building legacy of the Ghats.

The less popular Malnad offers a brief look into the day to day existences of its inhabitants, winding around a social story that supplements the normal miracles of the Western Ghats.

The investigation of the Western Ghats' less popular jewels uncovers an embroidery of stowed away cascades, unblemished biological systems, and energetic societies that improve the locale's story. From the isolated Gira Falls in Gujarat to the social gold mine of Malnad in Karnataka, each secret marvel adds a layer of intricacy to the Ghats' different and captivating story. As we celebrate both the notable and less popular wonders, let us treasure the unseen features that murmur stories of quietness, biodiversity, and social extravagance inside the hug of the Western Ghats.

Chapter 3

Northern Charms: Himalayan Cascades

Leaving on an excursion through the northern compasses of India uncovers a hypnotizing scene where transcending tops, thick woods, and flawless streams combine to make an embroidery of regular marvels. Among these charming highlights, the flowing cascades of the Himalayas stand as meaningful images of the area's excellence and greatness. These Himalayan fountains, settled in the midst of the great mountains, recount accounts of unending movement, environmental imperativeness, and social importance.

As we dig into the domain of Himalayan cascades, one can't neglect the notable excellence of Bhagsu Falls in Himachal Pradesh. Arranged close to the town of McLeod Ganj, Bhagsu Falls holds a respected spot in the hearts of the two local people and guests. The falls are taken care of by the spouting waters of the Bhagsu Bother stream, which starts from the Dhauladhar Reach.

What separates Bhagsu Falls isn't just its visual quality yet in addition the profound atmosphere that wraps the site. The cascade is named after the old Bhagsunath Sanctuary devoted to Master Shiva, which stands close by. Explorers and vacationers the same are attracted to the agreeable mix of normal excellence and strict importance. The flowing waters, outlined by the setting of rich plant life and the Himalayan pinnacles, make a quiet climate that welcomes thought and reflection.

As we move toward the east into the territory of Uttarakhand, the unlikely treasure of Kempty Falls unfurls in the beautiful valley of Mussoorie. Settled in the midst of the verdant slopes of the Garhwal Himalayas, Kempty Falls gets its name from the expression "camp-tea," as Britishers would arrange casual get-togethers here during the pilgrim time. The falls overflow down in numerous levels, shaping pools at different levels, where guests can submerge themselves in the cool, restoring waters.

Kempty Falls, however visited by sightseers, holds an ethereal beguile that orchestrates with its regular environmental factors. The flowing waters, in the midst of the setting of thick woods and fog covered mountains, make a visual scene that catches the pith of the Himalayan scene. The openness of Kempty Falls makes it a famous objective for those looking for reprieve from the late spring heat, and the falls become a declaration to the immortal charm of Himalayan conduits.

Wandering further into the Himalayan region, the great Tapovan Cascade in Gangotri Public Park uncovers itself as a secret gem. Taken care of by the Bhagirathi Waterway, the cascade slips from the Gangotri Ice sheet, encompassed by the perfect wild of the Garhwal Himalayas. Tapovan, meaning "woodland of contemplation," encapsulates the otherworldly holiness of the Himalayas, and the falls become a figurative excursion into the core of this hallowed domain.

Tapovan Cascade isn't simply a visual exhibition however a demonstration of the unique cycles molding the Himalayan scene. The icy mass took care of waters represent the soul of the district, supporting assorted environments and giving food to the networks downstream. The falls, concealed inside the folds of the mountains, act as a wake up call of the fragile equilibrium that characterizes the Himalayan biological system — an equilibrium that is unpredictably associated with the enduring progression of its streams.

Moving upper east into the territory of Sikkim, the less popular Seven Sisters Cascade unfurls as an outpouring of ethereal magnificence. Named after the seven northeastern territories of India, this cascade is situated close to the town of Gangtok. The Seven Sisters Cascade addresses the union of seven individual streams, each adding to the aggregate stream that slides from an extraordinary level.

What recognizes the Seven Sisters Cascade is its sheer scale and the vivid experience it offers to the people who stand in its presence. The flowing waters, outlined by the thick woods of Sikkim, make an exhibition that encapsulates the crude power and excellence of the Himalayas. The falls become an image of solidarity, as the seven streams converge into a solitary power — an illustration for the different yet interconnected societies of the northeastern states.

As we explore the northern edges of the Himalayas, the Nuranang Falls in Arunachal Pradesh remains as a demonstration of the district's tough excellence. Otherwise called the Bong Falls, Nuranang Falls is taken care of by the waters of the Tawang Stream, plummeting from the levels of the Eastern Himalayas. The falls, encompassed by snow-clad pinnacles and high knolls, encapsulate the crude and untamed charm of the Himalayan wild.

The social meaning of Nuranang Falls is entwined with the close by Tawang Religious community, quite possibly of the biggest cloister in India. Explorers and voyagers the same are attracted to the falls as they advance toward this sacrosanct site. The flowing waters become a figurative excursion for those looking for profound comfort, reflecting the more extensive job that Himalayan cascades play in the social texture of the district.

Venturing into the northernmost stretches of the Indian Himalayas, the entrancing Nohkalikai Falls in Meghalaya reveals itself as a stunning exhibition. Plunging from a level of more than 1,100 feet, Nohkalikai is the tallest dive cascade in India. Arranged close to the town of Cherrapunji, eminent for its storm rains, the falls get their name from an unfortunate nearby legend including a lady named Likai.

The sheer level and power of Nohkalikai Falls make a remarkable scene against the background of the Khasi Slopes. The water, diving into a turquoise pool underneath, structures a foggy cloak that adds to the falls' persona. The lavish scenes encompassing Nohkalikai Falls, portrayed by thick timberlands and lively vegetation, add to the general loftiness of the site. The falls become a demonstration of the geographical wonders that shape the Himalayas, as well as the social stories woven into the texture of the district.

In the northeastern province of Assam, the less popular Akashiganga Cascade close to Nameri Public Park offers a peaceful retreat in the midst of the Himalayan lower regions. Taken care of by the Jia Bhoroli Stream, the falls are accepted to be related with Hindu folklore, making them a consecrated site for explorers. The flowing waters, set against the background of the Eastern Himalayas, make a quiet climate that resounds with the otherworldly and normal substance of the locale.

Akashiganga Cascade represents the conversion of social love and natural importance that describes numerous Himalayan cascades. The encompassing scenes, set apart by different plant and creature life, add to the falls' charm as a safe-haven for biodiversity.

The musical progression of the waters turns into an image of congruity, interfacing the profound customs of the locale with the never-ending movement of the Himalayan streams.

As we investigate the appeal of Himalayan fountains, the spellbinding Khir Ganga Cascade in Himachal Pradesh uncovers itself as an unlikely treasure. Arranged in the Parvati Valley, close to the journey site of Khir Ganga, the cascade turns into a point of convergence for travelers and profound searchers the same. The journey to Khir Ganga takes travelers through high glades, thick timberlands, and all encompassing vistas, coming full circle in the hug of the flowing waters.

Khir Ganga Cascade, took care of by the waters of the Parvati Waterway, mirrors the unblemished magnificence of the Himalayan scenes. The falls, encompassed by transcending pinnacles and snow-clad highest points, become a safe-haven where voyagers can submerge themselves in the crude grandness of the district. The profound meaning of Khir Ganga adds an extraordinary aspect to the cascade insight, stressing the Himalayas' job as a material where nature and culture combine.

In the core of Jammu and Kashmir, the captivating Aharbal Cascade uncovers itself as a flowing magnum opus. Taken care of by the Veshu Waterway, Aharbal Falls drops from the Pir Panjal Reach, making a progression of ventured overflows that dazzle the onlooker. The falls, encompassed by snow capped knolls and thick woodlands, encapsulate the pristine greatness of the Himalayas.

Aharbal Cascade, frequently alluded to as the "Niagara of Kashmir," remains as a demonstration of the different scenes that characterize the locale. The thundering waters, outlined by snow-clad pinnacles and verdant valleys, become an image of the Himalayas' dynamic powers. The openness of Aharbal makes it a famous location for nature fans, offering a brief look into the perpetual excellence of the Himalayan streams.

Venturing into the lap of the Eastern Himalayas, the tranquil and less popular Borra Jhar Cascade in West Bengal adds a hint of persona to the district. Arranged close to the town of Darjeeling, this unexpected, yet invaluable treasure is taken care of by the Teesta Stream, wandering through the undulating scenes of the Himalayan lower regions. The falls, however not generally so broadly celebrated as a portion of their partners, offer a calm retreat for those looking for isolation in nature's hug.

Borra Jhar Cascade, encompassed by thick backwoods and tea manors, turns into a safe-haven where the orchestra of falling waters blends with the stirring leaves and the calls of avian inhabitants. The falls, concealed inside the folds of the Eastern Himalayas, mirror the untamed excellence of the area. Borra Jhar, in the same way as other Himalayan cascades, represents the sensitive harmony between human presence and the natural essentialness that supports the Eastern Himalayas.

The investigation of Himalayan fountains reveals a different and captivating story that traverses the length and broadness of the district. From the notable Bhagsu Falls in Himachal Pradesh to the unlikely treasures like Tapovan Cascade in Uttarakhand and Nohkalikai Falls in Meghalaya, each outpouring recounts an account of unending movement, social importance, and environmental imperativeness. The Himalayan cascades, outlined by the lofty pinnacles and rich scenes, become demonstrations of the persevering through magnificence of this unrivaled mountain range — a wonder that rises above both physical and otherworldly aspects.

3.1 Journey to the northern regions of India and explore the cascades nestled in the Himalayan foothills.

Setting out on an excursion toward the northern districts of India offers an enrapturing investigation of the Himalayan lower regions, where flowing cascades decorate the scene, making an orchestra of nature's greatness. As we cross the changed territory of this district, we experience notorious falls that reverberate with social importance, unlikely treasures that murmur stories of immaculate magnificence, and the cadenced progression of conduits that characterize the immortal appeal of the Himalayas.

In the core of Himachal Pradesh, Bhagsu Falls remains as a famous demonstration of the district's regular quality. Settled close to the town of McLeod Ganj, this cascade gets its name from the antiquated Bhagsunath Sanctuary committed to Ruler Shiva. Taken care of by the Bhagsu Bother stream, the falls overflow nimbly down rough slants, making a beautiful scene against the background of the Dhauladhar Reach.

What separates Bhagsu Falls isn't only its visual appeal yet in addition the other-worldly reverberation that envelopes the site. Pioneers and explorers are attracted to the amicable mix of regular excellence and strict sacredness. The flowing waters, in the midst of the rich plant life of the Himalayan scene, make a quiet climate that welcomes thought. The falls become a figurative excursion, where the actual drop of water reflects the otherworldly climb looked for by the individuals who visit this sacrosanct site.

As we adventure toward the east into Uttarakhand, Kempty Falls in Mussoorie unfurls as an unlikely treasure in the midst of the Garhwal Himalayas. The falls, settled in a beautiful valley, are taken care of by the spouting waters of the Kempty Stream. The multi-layered plummet of the falls structures pools at different levels, offering guests an opportunity to submerge themselves in the cool, reviving waters.

Kempty Falls, regardless of being a regularly visited vacationer location, holds an ethereal beguile that blends with its normal environmental factors. The flowing waters, set against the scenery of fog covered mountains, make a visual exhibition that catches the embodiment of the Himalayan scene.

The openness of Kempty Falls makes it a famous objective for those looking for reprieve from the late spring heat, and the falls stand as a demonstration of the immortal charm of Himalayan conduits.

Wandering further into the Himalayan breadth, Tapovan Cascade in Gangotri Public Park uncovers itself as a secret gem. Taken care of by the Bhagirathi Stream, the cascade drops from the Gangotri Icy mass, encompassed by the flawless wild of the Garhwal Himalayas. The name "Tapovan," meaning "woods of contemplation," typifies the profound holiness of the Himalayas, and the falls become a figurative excursion into the core of this sacrosanct domain.

Tapovan Cascade isn't simply a visual display yet a demonstration of the unique cycles forming the Himalayan scene. The ice sheet took care of waters represent the soul of the district, supporting different biological systems and giving food to networks downstream. The falls, concealed inside the folds of the mountains, act as a wake up call of the fragile equilibrium that characterizes the Himalayan environment — an equilibrium complicatedly associated with the lasting progression of its waterways.

Moving upper east into Sikkim, the Seven Sisters Cascade remains as an image of solidarity and regular overflow. Situated close to Gangtok, this cascade addresses the combination of seven individual streams, each adding to the aggregate stream that plunges from an incredible level. The falls, set against the scenery of the Eastern Himalayas, make an exhibition that typifies the crude power and excellence of the locale.

What recognizes the Seven Sisters Cascade is its sheer scale and the vivid experience it offers to the people who stand in its presence. The flowing waters, in the midst of the thick woodlands of Sikkim, make a scene that catches the crude power and magnificence of the Himalayas. The falls become a similitude for the different yet interconnected societies of the northeastern states, reverberating with the soul of solidarity in variety.

As we explore the eastern edges of the Himalayas, the less popular Nuranang Falls in Arunachal Pradesh spellbinds with its rough magnificence. Otherwise called the Bong Falls, Nuranang Falls is taken care of by the waters of the Tawang Waterway, dropping from the levels of the Eastern Himalayas. The falls, encompassed by snow-clad pinnacles and high knolls, typify the untamed appeal of the Himalayan wild.

The social meaning of Nuranang Falls is entwined with the close by Tawang Religious community, quite possibly of the biggest cloister in India. Explorers and voyagers are attracted to the falls as they advance toward this holy site. The flowing waters become a figurative excursion for those looking for profound comfort, reflecting the more extensive job that Himalayan cascades play in the social texture of the locale.

Venturing into the northernmost stretches of India, Nohkalikai Falls in Meghalaya unfurls as a stunning display. Plunging from a level of more than 1,100 feet, Nohkalikai is the tallest dive cascade in India. Arranged close to the town of Cherrapunji, eminent for its storm rains, the falls get their name from a heartbreaking nearby legend including a lady named Likai.

The sheer level and power of Nohkalikai Falls make a remarkable scene against the setting of the Khasi Slopes. The water, diving into a turquoise pool underneath, structures a cloudy shroud that adds to the falls' persona. The rich scenes encompassing Nohkalikai Falls, portrayed by thick woodlands and lively greenery, add to the general glory of the site. The falls become a demonstration of the land wonders that shape the Himalayas, as well as the social stories woven into the texture of the locale.

In Assam, the less popular Akashiganga Cascade close to Nameri Public Park offers a tranquil retreat in the midst of the Himalayan lower regions. Taken care of by the Jia Bhoroli Stream, the falls are accepted to be related with Hindu folklore, making them a holy site for explorers. The flowing waters, set against the scenery of the Eastern Himalayas, make a quiet air that resounds with the profound and normal embodiment of the district.

Akashiganga Cascade embodies the intersection of social worship and environmental importance that portrays numerous Himalayan cascades. The encompassing scenes, set apart by different plant and creature life, add to the falls' charm as a safe-haven for biodiversity. The musical progression of the waters turns into an image of congruity, interfacing the profound practices of the district with the never-ending movement of the Himalayan waterways.

As we investigate the appeal of Himalayan fountains, the spellbinding Khir Ganga Cascade in Himachal Pradesh uncovers itself as an unlikely treasure. Arranged in the Parvati Valley, close to the journey site of Khir Ganga, the cascade turns into a point of convergence for travelers and profound searchers the same. The trip to Khir Ganga takes swashbucklers through snow capped glades, thick woods, and all encompassing vistas, coming full circle in the hug of the flowing waters.

Khir Ganga Cascade, took care of by the waters of the Parvati Stream, mirrors the flawless excellence of the Himalayan scenes. The falls, encompassed by transcending pinnacles and snow-clad culminations, become a safe-haven where explorers can drench themselves in the crude grandness of the district. The profound meaning of Khir Ganga adds an otherworldly aspect to the cascade insight, underlining the Himalayas' job as a material where nature and culture unite.

In the core of Jammu and Kashmir, the charming Aharbal Cascade reveals itself as a flowing work of art. Taken care of by the Veshu Stream, Aharbal Falls slides from the Pir Panjal Reach, making a progression of ventured overflows that charm the spectator. The falls, encompassed by snow capped glades and thick backwoods, embody the pristine loftiness of the Himalayas.

Aharbal Cascade, frequently alluded to as the "Niagara of Kashmir," remains as a demonstration of the different scenes that characterize the locale. The thundering waters, outlined by snow-clad pinnacles and verdant valleys, become an image of the Himalayas' dynamic powers. The openness of Aharbal makes it a famous location for nature fans, offering a brief look into the perpetual magnificence of the Himalayan streams.

Venturing into the lap of the Eastern Himalayas, the quiet and less popular Borra Jhar Cascade in West Bengal adds a hint of persona to the locale. Arranged close to the town of Darjeeling, this unlikely treasure is taken care of by the Teesta Waterway, wandering through the undulating scenes of the Himalayan lower regions. The falls, however not generally so broadly celebrated as a portion of their partners, offer a tranquil retreat for those looking for isolation in nature's hug.

Borra Jhar Cascade, encompassed by thick woodlands and tea manors, turns into a safe-haven where the orchestra of falling waters blends with the stirring leaves and the calls of avian occupants. The falls, concealed inside the folds of the Eastern Himalayas, mirror the untamed magnificence of the area. Borra Jhar, in the same way as other Himalayan cascades, represents the fragile harmony between human presence and the natural imperativeness that supports the Eastern Himalayas.

As we explore the different scenes of the Himalayan lower regions, the flowing miracles become narrators, winding around stories of culture, otherworldliness, and biological lavishness. The excursion to these northern districts of India is a journey into the core of the Himalayas — a domain where every cascade turns into a section in the unfurling story of this glorious mountain range. From the famous Bhagsu Falls in Himachal Pradesh to the unlikely treasures like Borra Jhar Cascade in West Bengal, the Himalayan lower regions stand as a demonstration of the persevering through charm of flowing streams against the setting of transcending tops and untamed wild.

3.2 Discuss the spiritual and cultural importance of these waterfalls.

The flowing cascades settled in the Himalayan lower regions of northern India rise above their actual magnificence to become sacrosanct vaults of profound and social importance. These normal miracles, past their amazing visual allure, are entwined into the embroidery of strict convictions, antiquated fantasies, and social practices that have molded the existences of networks staying in the shadow of these grand pinnacles.

Bhagsu Falls in Himachal Pradesh, with its ethereal magnificence, holds profound otherworldly roots. Named after the Bhagsunath Sanctuary devoted to Ruler Shiva, the cascade turns into an emblematic connection between the earthbound and the heavenly. Pioneers, looking for otherworldly comfort, frequently embrace an excursion to Bhagsu Falls as a feature of their visit to the sanctuary.

The flowing waters, moving from the Dhauladhar Reach, are accepted to convey the heavenly endowments of Master Shiva, and a plunge in these consecrated waters is viewed as decontaminating.

The otherworldly excursion at Bhagsu Falls stretches out past the sanctuary, as guests are encompassed in the serenity of the Himalayan scene. The cadenced sound of falling water turns into a thoughtful background, welcoming reflection and examination. The falls, outlined by lavish vegetation and fog covered tops, make an air where nature and otherworldliness merge, offering a safe-haven for those looking for fellowship with the heavenly.

Moving toward the east into Uttarakhand, the Kempty Falls in Mussoorie becomes a grand fascination as well as a social retreat. The falls, set in the midst of the Garhwal Himalayas, are saturated with verifiable importance. During the pilgrim period, Britishers would put together casual get-togethers at Kempty Falls, and the actual name, got from "camp-tea," mirrors this frontier association. The falls, encompassed by thick backwoods and fog loaded mountains, summon a feeling of wistfulness for a period gone by.

Kempty Falls, while having embraced its pilgrim past, likewise holds social pertinence for the neighborhood networks. The openness of the falls makes it a well known objective for the two travelers and explorers. Guests frequently take part in formal works on, offering petitions and performing functions along the banks of the flowing waters. Kempty Falls, in its double job as a social and verifiable milestone, turns into an extension associating at various times in the Himalayan scene.

In the core of Gangotri Public Park, the Tapovan Cascade remains as a demonstration of the otherworldly sacredness of the locale. Taken care of by the Bhagirathi Stream and slipping from the Gangotri Glacial mass, Tapovan epitomizes the embodiment of a "timberland of reflection." The Himalayas, worshipped as the house of the divine beings, are scratched into the otherworldly cognizance of individuals. The flowing waters of Tapovan are accepted to convey the virtue and favors of the Gangotri, thought about the wellspring of the hallowed Ganges Stream.

For travelers undertaking the exhausting excursion to Gangotri, Tapovan Cascade turns into a holy waypoint — a spot to reflect, revive, and look for profound illumination. The falls, concealed inside the folds of the Garhwal Himalayas, are wrapped in an atmosphere of heavenly energy. The flowing waters, cool and unadulterated, represent the hallowed excursion of the Ganges, from its frigid beginnings to the fields where it gives life and endowments.

Venturing into Sikkim, the Seven Sisters Cascade unfurls as a sign of social solidarity and regular overflow. Named after the seven northeastern provinces of India, this cascade turns into an image of the interconnectedness of different societies.

The northeastern district, frequently alluded to as the "Seven Sisters," is known for its rich social woven artwork and ethnic variety. The falls, with seven individual streams meeting into a particular power, reflect the solidarity in variety that describes individuals of this locale.

For the nearby networks in Sikkim, the Seven Sisters Cascade isn't simply a scene yet a social symbol. The flowing waters are related with customary convictions and ceremonies, and the site frequently turns into a background for celebrations and functions. The falls, outlined by the Eastern Himalayas, become a material whereupon the social personality of the district is painted — a demonstration of the persevering through soul of solidarity notwithstanding variety.

In Arunachal Pradesh, Nuranang Falls, otherwise called Bong Falls, holds social importance entwined with the otherworldly texture of the Eastern Himalayas. Taken care of by the waters of the Tawang Waterway, the falls are arranged in nearness to the Tawang Cloister, a venerated Buddhist site. The excursion to Nuranang Falls turns into a journey for the two local people and guests, as the flowing waters are accepted to convey the favors of the religious community and the insight of old lessons.

The social reverberation of Nuranang Falls is enhanced by the fables and legends that encompass the district. The falls, encompassed by snow-clad pinnacles and high knolls, become a course for narrating — a medium through which the oral customs of the Himalayas are saved. The social embroidery of Arunachal Pradesh, with its lively tones, tracks down articulation in the great drop of the falls.

Moving to Meghalaya, Nohkalikai Falls uncovers itself as an image of social stories woven into the scene. Plunging from a level of north of 1,100 feet, Nohkalikai is the tallest dive cascade in India. The falls, arranged close to Cherrapunji, a spot famous for its storm downpours, get their name from a neighborhood legend including a lady named Likai.

The social significance of Nohkalikai Falls is interwoven with the heartbreaking story of Likai, whose name signifies "bounce of Likai." As per the legend, Likai, suffering from extreme melancholy after the demise of her little girl, dove into the void close to the falls. The powerful story turns into a piece of the oral customs of the Khasi public, adding a layer of social profundity to the normal scene of Nohkalikai.

In Assam, the Akashiganga Cascade close to Nameri Public Park combines social adoration with biological importance. Taken care of by the Jia Bhoroli Waterway, the falls are accepted to be related with Hindu folklore. As indicated by nearby legends, the cascade is viewed as a sign of the tears of Master Shiva, shedding sorrow over the deficiency of his partner Sati. Travelers and fans visit Akashiganga to look for favors and proposition petitions, transforming the site into a holy safe-haven.

The social significance of Akashiganga Cascade stretches out past strict convictions to envelop the rich biodiversity of the Eastern Himalayas. The encompassing scenes, set apart by thick woods and different widely varied vegetation, become a demonstration of the interconnectedness of culture and nature. The falls, set against the scenery of the Himalayan lower regions, become where otherworldly adoration and environmental preservation merge.

In Himachal Pradesh, the Khir Ganga Cascade, settled in the Parvati Valley, adds an otherworldly and supernatural aspect to the Himalayan scene. The cascade is arranged close to the journey site of Khir Ganga, where, as per Hindu folklore, Ruler Shiva is accepted to have thought. The trip to Khir Ganga turns into a profound excursion, driving voyagers through high glades and thick woodlands to the flowing waters.

Khir Ganga Cascade, took care of by the Parvati Stream, turns into a safe house for those looking for profound comfort and fellowship with nature. The falls, encompassed by transcending pinnacles and snow-clad highest points, epitomize the crude magnificence of the Himalayas. The profound meaning of Khir Ganga turns into a point of convergence for explorers and searchers, mirroring the immortal association between the physical and otherworldly domains in the Himalayan scene.

In Jammu and Kashmir, the Aharbal Cascade unfurls as a flowing show-stopper with both social and environmental importance. Taken care of by the Veshu Waterway, Aharbal Falls slides from the Pir Panjal Reach, making a progression of ventured overflows. The falls, frequently alluded to as the "Niagara of Kashmir," become a social milestone, drawing the two local people and sightseers to encounter their loftiness.

Aharbal Cascade holds social significance as a site for celebrations and get-togethers, where networks meet up to praise the regular abundance of the district. The falls, encompassed by elevated glades and thick timberlands, become an image of the untainted magnificence of the Himalayas. Aharbal remains as a demonstration of the sensitive harmony between human presence and the biological essentialness that supports the Himalayan lower regions.

In West Bengal, the Borra Jhar Cascade close to Darjeeling adds a hint of persona toward the Eastern Himalayas. Taken care of by the Teesta Stream, the falls are arranged in the midst of undulating scenes and tea estates. While not quite as broadly celebrated as a portion of its partners, Borra Jhar Cascade turns into a tranquil retreat for those looking for isolation and contemplation.

The falls, concealed inside the folds of the Eastern Himalayas, become a social safe-haven where the ensemble of falling waters blends with the stirring leaves and the calls of avian occupants. Borra Jhar, in the same way as other Himalayan cascades, represents the fragile harmony between human presence and the natural imperativeness that supports the area.

The falls, encompassed by thick timberlands and tea manors, become a material where nature and culture unite in a quiet hug.

As we dig into the profound and social significance of these Himalayan cascades, the more extensive story unfurls — an account that traverses hundreds of years, envelops different networks, and interweaves with the biological imperativeness of the locale. Every cascade, whether celebrated like Bhagsu Falls or secret like Borra Jhar, turns into a social reference point, enlightening the rich embroidery of convictions, customs, and practices that have molded the existences of the people who call the Himalayan lower regions home.

The otherworldly meaning of these cascades is well established in the strict convictions of individuals, where the flowing waters are frequently viewed as consecrated and saturated with divine energy. The Himalayas, respected as the home of divine beings and goddesses in Hindu folklore, give a setting against which these cascades become residing signs of the consecrated. Pioneers set out on excursions to these falls, to observe their normal excellence as well as to participate in ceremonies that associate them to the heavenly powers accepted to dwell in these flawless scenes.

Socially, these cascades act as living landmarks to the legacy of the networks staying in their area. The stories and legends related with the falls, went down through ages, become a basic piece of the neighborhood fables. Whether it's the shocking legend of Likai at Nohkalikai Falls or the frontier time casual get-togethers at Kempty Falls, every story adds to the social character of the area, adding layers of importance to the flowing waters.

The celebrations and services held at these cascades further highlight their social significance. From strict festivals at Akashiganga to common get-togethers at Aharbal, these occasions become events for individuals to meet up, building up a feeling of local area and shared legacy. The falls, outlined by the transcending pinnacles of the Himalayas, become regular amphitheaters where social articulations unfurl against the scenery of perfect nature.

Also, the environmental meaning of these cascades couldn't possibly be more significant. The Himalayan lower regions, home to assorted greenery, track down food in the lasting progression of these conduits. The falls add to the general wellbeing of the environment, giving water to both natural life and human networks downstream. The fragile harmony between social practices and biological protection becomes clear, featuring the entwined idea of human existence with the normal world.

The profound and social significance of the Himalayan cascades rises above their actual excellence. Bhagsu Falls, Kempty Falls, Tapovan Cascade, Seven Sisters Cascade, Nuranang Falls, Nohkalikai Falls, Akashiganga Cascade, Khir Ganga Cascade,

Aharbal Cascade, and Borra Jhar Cascade — all weave a story that mirrors the profound otherworldly association among mankind and the Himalayas.

These cascades are not just topographical highlights; they are sacrosanct milestones, social standards, and biological life savers that encapsulate the pith of the Himalayan lower regions — a district where nature, culture, and otherworldliness meet in an agreeable dance.

3.3 Share stories of adventure and exploration in the challenging terrains.

Leaving on an endeavor into the difficult territories of the Himalayan lower regions is an endeavor that entices swashbucklers to investigate the crude magnificence of nature, face their own constraints, and disentangle the untold stories concealed inside the folds of these considerable scenes. Each step into this rough territory is a demonstration of human versatility, a discourse with the components, and a potential chance to cut accounts of experience and investigation that reverberation through the ages.

The excursion to Bhagsu Falls in Himachal Pradesh, settled close to the town of McLeod Ganj, starts with a journey that navigates rough slants and thick backwoods. The Himalayan lower regions, with their capricious climate and steep risings, set up for a difficult yet remunerating experience. As travelers explore the restricted paths, the far off thunder of Bhagsu Falls turns into a directing tune, attracting them nearer to the flowing waters.

The trip to Bhagsu Falls isn't simply an actual undertaking yet an otherworldly odyssey. Explorers, globe-trotters, and nature fans join on this way, looking for comfort in the immaculate scenes that unfurl en route. The excursion turns into an investigation of self-disclosure, with each diversion uncovering new vistas and unanticipated difficulties. As travelers at long last substitute the presence of Bhagsu Falls, the sheer greatness of the regular scene turns into a summit of their gutsy soul and assurance to overcome the Himalayan paths.

Wandering toward the east into Uttarakhand, the way to Kempty Falls in Mussoorie presents a remarkable mix of experience and verifiable reverberation. The journey envelops undulating territories, thick timberlands, and intermittent looks at the snow-covered Garhwal Himalayas. The verifiable association of Kempty Falls, when visited by pilgrim period casual get-togethers, adds a layer of interest to the excursion, welcoming voyagers to interface with the past.

The test lies in exploring the different scenes as well as in translating the narratives carved into the slopes. As travelers track the paths to Kempty Falls, the air becomes loaded down with the fragrance of pine, and the far off hints of the flowing waters allure them forward. The falls, settled in a valley encompassed by fog loaded mountains, become a compensation for the people who set out to wander into the core of the Himalayan lower regions.

In Gangotri Public Park, the journey to Tapovan Cascade unfurls as an odyssey into the core of the Garhwal Himalayas.

The difficult landscape, set apart by frigid moraines and rough bluffs, requests actual perseverance as well as a profound association with the otherworldly embodiment of the locale. The Bhagirathi Stream, starting from the Gangotri Glacial mass, turns into a friend on this excursion, directing travelers toward the flowing waters of Tapovan.

The experience to Tapovan is as much about vanquishing the geological difficulties for what it's worth about giving up to the antiquated energy of the Himalayas. Travelers experience snow capped knolls, intriguing Himalayan vegetation, and the always present feeling of rise that describes the area. As they climb to Tapovan, the falls uncover themselves as a great scene, outlined by transcending tops and the ethereal excellence of chilly scenes. The excursion turns into a fellowship with nature, where the soul of investigation converges with the immortal charm of the Himalayan wild.

Investigating Sikkim discloses the experience of arriving at the Seven Sisters Cascade close to Gangtok. The paths wander through thick woodlands, offering looks at different verdure. The climb is steep, and the territory presents both physical and mental difficulties. The falls, named after the seven northeastern states, become an image of solidarity in the midst of the difficulties looked by travelers exploring the mind boggling pathways.

As explorers advance toward the Seven Sisters Cascade, they cross the different scenes of Sikkim — the place that is known for religious communities, rhododendron-filled valleys, and all encompassing perspectives on the Eastern Himalayas. The trip turns into a drenching into the social variety of the area, with experiences with nearby networks and their stories of versatility. The falls, outlined by the foggy pinnacles, become a signal for those looking for the excitement of investigation in the midst of the rich embroidery of Sikkimese scenes.

Venturing into the Eastern Himalayas, the journey to Nuranang Falls in Arunachal Pradesh unfurls as a campaign into strange regions. The locale, set apart by far off towns and unblemished wild, provokes globe-trotters to explore through thick backwoods and steep slopes. The excursion turns into an orchestra of sounds — the stirring leaves, the calls of colorful birds, and the far off thunder of the flowing waters.

Nuranang Falls, otherwise called Bong Falls, isn't simply an objective; it is a story ready to be found. Travelers become narrators as they explore the tough scenes, experiencing the social subtleties of Arunachal Pradesh. The falls, encompassed by snow-clad pinnacles and high knolls, become a point of convergence for stories of old customs, fables, and the getting through soul of investigation in the Eastern Himalayas.

Moving to Meghalaya, the journey to Nohkalikai Falls close to Cherrapunji unfurls as an emotional adventure in a land known for its storm downpours and lavish scenes.

The paths wind through verdant valleys, testing the travelers with tricky ways and the always present shade of fog. Nohkalikai, the tallest dive cascade in India, turns into a guide noticeable from a far distance, drawing travelers into the core of the Khasi Slopes.

The experience to Nohkalikai is a hit the dance floor with the components — the cadenced patter of downpour, the rich fragrance of wet earth, and the far off reverberations of the falls. Adventurers explore the tricky rocks, their faculties elevated by the steady mumble of falling water. As they stand at the edge of the incline, sitting above the turquoise pool underneath, the sheer power and magnificence of Nohkalikai become a permanent memory scratched into the courageous soul.

In Assam, the trip to Akashiganga Cascade close to Nameri Public Park turns into an excursion into the otherworldly and regular domains of the Eastern Himalayas. The difficult territory, set apart by thick backwoods and waterway intersections, requests an association with the flawless wild. Akashiganga, related with Hindu folklore and considered an indication of Ruler Shiva's tears, turns into a journey for both the passionate and the brave.

As travelers explore the paths to Akashiganga, they experience the biodiversity of Nameri Public Park — a safe-haven for elephants, tigers, and various avian species. The experience turns into an agreeable mix of social veneration and natural investigation. The falls, encompassed by the Eastern Himalayan scenes, become a combination point for those looking for both profound comfort and the excitement of exploring through untamed nature.

In Himachal Pradesh, the trip to Khir Ganga Cascade in the Parvati Valley rises above the limits of an actual excursion. The paths wind through high knolls, natural aquifers, and all encompassing vistas of the Himalayan reaches. Khir Ganga, accepted to be the site where Master Shiva reflected, turns into an objective that entices profound searchers and swashbucklers the same.

The experience to Khir Ganga unfurls as a journey, with travelers navigating old ways trampled by explorers for a really long time. The falls, took care of by the Parvati Stream, become a position of reflection and greatness. As travelers drench themselves in the normal natural aquifers and witness the flowing waters, they become piece of a continuum — a deep rooted custom of investigation and otherworldly journey in the Himalayan lower regions.

In Jammu and Kashmir, the journey to Aharbal Cascade in the Pir Panjal Reach turns into an odyssey into the pristine excellence of the Kashmir Valley. The paths wind through elevated glades, pine backwoods, and broad vistas of snow-clad pinnacles. Aharbal, frequently alluded to as the "Niagara of Kashmir," turns into an objective that epitomizes the magnificence of the Western Himalayas.

The experience to Aharbal is a discourse with the different scenes that characterize the Kashmir Valley. Adventurers experience nearby networks, their conventional practices, and the always changing tones of the view. The falls, with their ventured fountains and thundering waters, become a characteristic amphitheater where the experience unfurls against the background of perfect wild.

Traveling into West Bengal, the journey to Borra Jhar Cascade close to Darjeeling turns into an endeavor into the lesser-investigated scenes of the Eastern Himalayas. The paths wind through tea estates, thick woods, and undulating landscapes. Borra Jhar, concealed inside the folds of the Eastern Himalayas, turns into a retreat for those looking for isolation and the excitement of finding the unknown.

The experience to Borra Jhar is a tangible encounter — the smell of tea leaves, the stirring of leaves, and the far off reverberations of falling water. Adventurers explore through the maze of tea domains, experiencing the different vegetation that portray the Eastern Himalayan lower regions. The falls, however not as broadly celebrated, become a mysterious safe house for the individuals who try to wander into the core of the less popular scenes.

As we wind through these stories of experience and investigation in the difficult territories of the Himalayan lower regions, the more extensive story arises — a story of strength, fellowship with nature, and the unstoppable soul of the people who set out to step where not many have wandered. Every cascade, whether Bhagsu Falls, Kempty Falls, Tapovan Cascade, Seven Sisters Cascade, Nuranang Falls, Nohkalikai Falls, Akashiganga Cascade, Khir Ganga Cascade, Aharbal Cascade, or Borra Jhar Cascade, turns into a section in the fantastic adventure of human investigation against the material of the considerable Himalayan scene.

Chapter 4

Eastern Mystique
Waterfalls in the East

Setting out on an excursion to investigate the cascades of the eastern locales of India discloses an embroidery of persona, where flowing waters become the strings winding around stories of normal marvels, social extravagance, and otherworldly importance. The Eastern Himalayas, embellished with lavish plant life, different scenes, and a heap of conduits, coax globe-trotters and lovers the same to dive into the charming domains of these cascades.

In the northeastern province of Sikkim, the popular Banjhakri Falls remains as an encapsulation of Eastern persona. Settled in the midst of thick backwoods and emerald scenes, this cascade holds social importance for the neighborhood people. The falls are named after the Banjhakri, customary healers in Sikkim, and visiting the site is accepted to achieve otherworldly recuperating.

Banjhakri Falls, with its shroud of fog and the mitigating sound of falling water, turns into a safe-haven where nature and otherworldliness join. As guests navigate the pathways prompting the falls, they are wrapped in the lively greenery of the Eastern Himalayas.

The falls, outlined by greenery covered rocks and verdant foliage, reverberation the mysterious vibe that characterizes Sikkim — a land where old practices and regular magnificence orchestrate.

Venturing into the thick backwoods of Meghalaya, the flowing excellence of Krang Suri Falls disentangles as an unlikely treasure inside the East. Arranged close to Jowai, this cascade epitomizes the immaculate charm of the district. The trip to Krang Suri turns into an endcavor through wandering paths, thick overhangs, and the musical hints of the streaming water.

Krang Suri Falls, encompassed by turquoise pools and limestone developments, enamors with its ethereal magnificence. The detached idea of the falls adds to the persona, making it an objective for those looking for comfort in the midst of the immaculate scenes of Meghalaya. The excursion to Krang Suri turns into a disclosure of Eastern persona — a mix of regular polish and the quietude that characterizes the less-investigated corners of the Eastern Himalayas.

Wandering into Arunachal Pradesh, the supernatural charm of Bap Teng Kang Cascade coaxes swashbucklers. Arranged close to Tawang, this cascade is settled in the midst of elevated glades and snow-clad pinnacles, making a dreamlike scenery for investigation. The excursion to Bap Teng Kang includes exploring through immaculate scenes, where the quietness is just broken by the flowing waters and the murmurs of the Himalayan breezes.

Bap Teng Kang Cascade, took care of by the icy waters of the Tawang Waterway, turns into an image of the crude excellence that describes the Eastern Himalayas. The falls, encompassed by the considerable pinnacles of the locale, disclose a feeling of serenity and secret. As travelers feel overwhelmed by Bap Teng Kang, they become piece of a story that rises above the actual domains — an experience with the enchanted heart of the Eastern Himalayas.

In Nagaland, the mysterious Dzükou Falls adds to the persona of the Eastern scenes. Concealed inside the Dzükou Valley, this cascade is important for a district known for its biodiversity and immaculate biological systems. The excursion to Dzükou Falls includes traveling across thick timberlands, crossing streams, and submerging in the rich verdure that characterizes the area.

Dzükou Falls, outlined by emerald slopes and dynamic blossoms, turns into a demonstration of the immaculate wild of Nagaland. The falls, with their delicate outpouring, reverberation the serenity of the Eastern Himalayas. As swashbucklers investigate the Dzükou Valley, they are welcomed by the amicable orchestra of nature — a tune that typifies the persona of Nagaland's secret cascades.

Moving to Mizoram, Vantawng Falls remains as a grand display inside the Eastern Himalayas. Plunging from the bluffs of the Jawline Slopes, this cascade turns into an image of Mizoram's regular loftiness. The excursion to Vantawng Falls includes exploring through the sloping territories, offering all encompassing perspectives on the encompassing scenes.

Vantawng Falls, with its multi-layered plunge and the thunder of flowing waters, catches the embodiment of Mizoram's persona. The falls are encircled by rich plant life, making a visual ensemble that resounds with the lively culture of the locale. As guests stand within the sight of Vantawng, they witness the combination of geology and culture — a demonstration of the Eastern Himalayas' capacity to enchant and move.

In Manipur, the hypnotizing Hundung Mangva Falls uncovers itself as a secret fortune inside the Eastern scenes. Arranged close to Ukhrul, this cascade is portrayed by unblemished environmental factors and the tranquility characterizes the district. The excursion to Hundung Mangva includes crossing through the slopes, offering looks at Manipur's undulating geography.

Hundung Mangva Falls, with its dropping waters and verdant environmental factors, turns into a position of break for those looking for the persona of Manipur. The falls, took care of by the Hundung Mangva Waterway, reflect the pristine excellence of the Eastern Himalayas. As globe-trotters investigate the pathways prompting Hundung Mangva, they become observers to the sensitive dance among nature and persona in the core of Manipur.

Diving into the scenes of Tripura, the captivating Sepahijala Cascade adds a bit of sorcery toward the Eastern Himalayas. Arranged inside the Sepahijala Untamed life Safe-haven, this cascade turns into a point of convergence for nature sweethearts and pilgrims. The excursion to Sepahijala includes navigating through the asylum's different biological systems, where untamed life and normal magnificence coincide.

Sepahijala Cascade, encompassed by thick backwoods and the lively greenery of Tripura, turns into a door to the enchanted domains of the Eastern Himalayas. The falls, with their delicate fountain and detached feeling, make a safe-haven inside the safe-haven. As guests drench themselves in the biodiversity of Sepahijala, they become piece of a story that commends the interconnectedness of life in the Eastern scenes.

In Assam, the enamoring Kakochang Cascade unfurls as an entrancing display inside the Eastern Himalayas. Arranged close to Jorhat, this cascade is settled in the midst of tea plants and undulating scenes. The excursion to Kakochang includes exploring through the pleasant tea domains, offering a brief look into the social embroidery of Assam.

Kakochang Cascade, with its layered plunge and the cadenced progression of water, turns into a position of reflection and investigation. The falls are encircled by the lavish vegetation of Assam, making a peaceful setting for globe-trotters.

As guests dig into the core of Kakochang, they are encompassed by the social subtleties and regular excellence that characterize the persona of the Eastern Himalayas.

Setting out on an investigation of the cascades in the Eastern Himalayas turns into an excursion into the obscure — a mission to unwind the secrets concealed inside the different scenes of Sikkim, Meghalaya, Arunachal Pradesh, Nagaland, Mizoram, Manipur, Tripura, and Assam. Every cascade, from Banjhakri Tumbles to Kakochang Cascade, turns into an entryway to a domain where nature, culture, and persona meet in an agreeable dance.

The persona of the Eastern Himalayan cascades goes past their actual magnificence; it is implanted in the social texture and profound reverberation of the areas they elegance. Banjhakri Falls, named after customary healers, becomes a characteristic scene as well as a position of profound mending — a nexus where the mysterious and the restorative entwine. Krang Suri Falls in Meghalaya, concealed inside immaculate scenes, welcomes swashbucklers to turn out to be essential for a story that praises the neglected corners of the East.

The cascades in Arunachal Pradesh, for example, Bap Teng Kang and Dzükou Falls in Nagaland, reverberation the untamed wild and profound holiness of the Eastern Himalayas. The investigation of these falls isn't simply an actual excursion however a fellowship with the old energy that penetrates the Eastern scenes. Vantawng Falls in Mizoram and Hundung Mangva Falls in Manipur stand as landmarks to the regular greatness that characterizes the persona of these locales.

In Tripura, Sepahijala Cascade turns into a passage to the interconnected biological systems of the Sepahijala Untamed life Safe-haven, offering a brief look into the fragile equilibrium of life in the Eastern Himalayas. Kakochang Cascade in Assam, encompassed by tea homes, turns into a social retreat where the orchestra of falling water blends with the rich legacy of the locale.

As globe-trotters explore the paths prompting these Eastern Himalayan cascades, they become narrators, winding around stories of investigation, social revelation, and profound association. The falls, with their flowing waters and verdant environmental elements, become sections in a bigger story — a tribute to the persona that characterizes the Eastern Himalayas. Each step into the obscure turns into a hit the dance floor with the supernatural, where the cascades act as guides into the core of a district that proceeds to enamor and captivate the people who set out to investigate its profundities.

The cascades of the Eastern Himalayas exemplify a persona that rises above the limits of geology and culture. Banjhakri Falls, Krang Suri Falls, Bap Teng Kang Cascade, Dzükou Falls, Vantawng Falls, Hundung Mangva Falls, Sepahijala Cascade, and Kakochang Cascade — all become entrances into a domain where the mysterious and the normal merge.

As travelers submerge themselves in the investigation of these falls, they become observers to the deep rooted stories, social subtleties, and profound resonances that characterize the baffling excellence of the Eastern Himalayas.

4.1 Uncover the lesser-explored waterfalls in the eastern part of India.

Setting out on an excursion to reveal the lesser-investigated cascades in the eastern piece of India discloses a gold mine of unlikely treasures, where nature's greatness meets the charm of strange regions. Past the all around trampled ways lie flows that have escaped the spotlight, standing by to enthrall pilgrims with their unblemished excellence and interesting stories.

In the northeastern territory of Arunachal Pradesh, the less-observed Nyngnging Cascade remains as a demonstration of the immaculate scenes of the locale. Concealed in the lavish plant life of the Eastern Himalayas, Nyngnging welcomes travelers to explore through thick woods and uneven territories to observe its tranquil fountain. The excursion to Nyngnging isn't simply an actual investigation however an endeavor into the core of Arunachal's crude wild.

Nyngnging Cascade, with its delicate plummet and the cadenced progression of water, epitomizes the nuance that describes numerous less popular falls. Encircled by the lively verdure of Arunachal Pradesh, the falls become a safe-haven for those looking for isolation and the excitement of revelation. As pilgrims stand within the sight of Nyngnging, they become piece of a chosen handful who have unwound the secret of this secret cascade.

Moving to the territory of Nagaland, the moderately dark Rangapahar Cascade allures fearless spirits to investigate its isolated excellence. Settled inside the verdant scenes close to Dimapur, the excursion to Rangapahar includes exploring through the undulating landscapes and thick vegetation that shroud this less popular pearl. The falls, however not as broadly perceived, hold their own appeal in the Eastern Himalayan embroidery.

Rangapahar Cascade, with its unblemished pools and greenery covered rocks, turns into a shelter for those looking for off in an unexpected direction undertakings. The falls, encompassed by the ensemble of nature, offer a retreat from the buzzing about of metropolitan life. As pioneers adventure into the quietude of Rangapahar, they become observers to the unpretentious yet charming magnificence that characterizes the lesser-investigated cascades of Nagaland.

Digging into the Eastern Ghats of Odisha reveals the secret wonder of Khandadhar Falls, a lesser-investigated overflow that graces the rough landscape close to Kendujhar. While not quite so generally recognized as a portion of its partners, Khandadhar enthralls with its transcending plunge and the untamed scenes that envelope it. The excursion to Khandadhar turns into a trip across thick woodlands and rough outcrops, offering looks at Odisha's different geology.

Khandadhar Falls, with its booming thunder and the shower of fog in the air, remains as an impressive exhibition in the Eastern Ghats. The falls, outlined by the rich biodiversity of the district, become an image of the untamed excellence that anticipates those ready to wander past the recognizable paths. As travelers stand at the foundation of Khandadhar, they become pioneers in unwinding the secrets covered inside the lesser-investigated cascades of Odisha.

Wandering into the core of Jharkhand, the less popular Lodh Falls arises as a secret gem in the midst of the thick backwoods close to Netarhat. While not as visited by sightseers, Lodh Falls coaxes those looking for a serene departure into nature's hug. The excursion to Lodh includes exploring through the wild, where the falls uncover themselves as a flowing marvel in the Chota Nagpur Level.

Lodh Falls, with its ventured plunge and the unblemished environmental factors, turns into a demonstration of the immaculate magnificence that describes the lesser-investigated cascades of Jharkhand. The falls, took care of by the Burha Waterway, make a quiet feel that resounds with the serenity of the Eastern scenes. As voyagers stand at the vantage focuses sitting above Lodh, they become pioneers in opening the mysteries of Jharkhand's less popular normal marvels.

In the northeastern territory of Tripura, the moderately dark Hattabari Cascade welcomes globe-trotters to reveal its appeal in the midst of the verdant scenes. Concealed in the slopes close to Amarpur, Hattabari stays off in an unexpected direction, giving an open door to adventurers to explore through the rich plant life and rough territories that lead to its disconnected safe house. The excursion to Hattabari turns into an introduction to the lesser-investigated cascades of Tripura.

Hattabari Cascade, with its smooth fountain and the serenity of its environmental elements, offers a reprieve from the traditional vacationer circuits. The falls, settled inside the Eastern Himalayan lower regions, become a location for those looking for regular excellence as well as the excitement of finding stowed away fortunes. As swashbucklers stand before Hattabari, they become pioneers in the investigation of Tripura's less popular jewels.

Setting out on the investigation of the lesser-investigated cascades in the eastern piece of India isn't just a mission for actual experience however an excursion into the immaculate domains of nature's abundance. Nyngnging Cascade in Arunachal Pradesh, Rangapahar Cascade in Nagaland, Khandadhar Falls in Odisha, Lodh Falls in Jharkhand, and Hattabari Cascade in Tripura — all become waypoints in a story that commends the excellence hid inside the Eastern scenes.

As travelers cross through the secret paths, they become observers to the nuance and beauty of these less popular falls. Nyngnging, with its delicate plummet, reflects the tranquil mood of Arunachal Pradesh's wild.

Rangapahar, settled in the quietude of Nagaland, welcomes adventurers to reveal its isolated appeal. Khandadhar, roaring in the Eastern Ghats of Odisha, turns into a demonstration of the crude force of nature. Lodh, flowing in the Chota Nagpur Level of Jharkhand, uncovers a ventured wonder in the midst of the wild. Hattabari, covered in the slopes of Tripura, offers a peaceful break for those ready to wander past the ordinary.

Generally, the lesser-investigated cascades in the eastern piece of India become passages to a domain where the traditional gives approach to the phenomenal. Each fountain, concealed in the folds of Arunachal Pradesh, Nagaland, Odisha, Jharkhand, and Tripura, turns into a part in the bigger story of investigation — a demonstration of the different and frequently ignored excellence that characterizes the Eastern scenes.

4.2 Discuss the unique features of these cascades and their impact on the local communities.

Diving into the extraordinary elements of the lesser-investigated cascades in the eastern piece of India reveals their actual magnificence as well as the significant effect they have on the neighborhood networks staying in their area. Past being pleasant scenes, these fountains assume a vital part in shaping the lives, customs, and societies of the districts they elegance. Every cascade, from Nyngnging in Arunachal Pradesh to Hattabari in Tripura, flaunts particular elements that add to their charm and importance.

Starting with Nyngnging Cascade in Arunachal Pradesh, the uniqueness of this fountain lies in its quiet and unblemished environmental elements. Concealed in the Eastern Himalayas, Nyngnging is described by a delicate plunge that bestows a feeling of serenity to its mood. The falls, encompassed by lavish plant life and energetic vegetation, make a characteristic safe-haven that entices those looking for comfort and an association with the immaculate wild.

The effect of Nyngnging on the neighborhood networks goes past its stylish allure. The falls act as a wellspring of profound motivation and social importance for the native clans of Arunachal Pradesh. The delicate fountain turns into an image of the sensitive harmony among people and nature, encouraging a feeling of veneration and concurrence. Neighborhood people group frequently integrate the falls into their old stories, crediting enchanted characteristics to the flowing waters, and services might be held to respect the soul of Nyngnging.

In Nagaland, Rangapahar Cascade uncovers its exceptional appeal in the midst of the undulating territories close to Dimapur. While not quite so generally perceived as a portion of its partners, Rangapahar stands apart with its segregated excellence. The falls, concealed inside thick vegetation, present an outpouring that converges with the regular forms of the scene. The remarkable element of Rangapahar lies in its capacity to offer a retreat from the rushing about, giving a quiet sanctuary to contemplation and association with nature.

The effect of Rangapahar on the neighborhood networks stretches out to its job as a characteristic safe-haven. The falls become a space for social get-togethers, where networks meet up to commend their association with nature. Rangapahar cultivates a feeling of ecological stewardship, empowering neighborhood drives for preservation and maintainable practices. The falls, settled inside the Nagaland wild, become a common image of local area character and an indication of the district's natural legacy.

Going to Odisha, Khandadhar Falls in the Eastern Ghats stands tall as a less popular miracle with its transcending drop and untamed scenes. The extraordinary component of Khandadhar lies in its considerable presence, as the waters plunge from incredible levels, making a striking display. The falls, encompassed by thick backwoods and rough outcrops, typify the crude force of nature in the Eastern Ghats.

The effect of Khandadhar on the nearby networks is entwined with its job as a characteristic milestone. The falls frequently become a point of convergence for comprehensive developments and celebrations, uniting networks in the hug of the Eastern Ghats. Khandadhar adds to the jobs of neighborhood occupants through the travel industry, setting out monetary open doors while likewise bringing issues to light about the significance of safeguarding the interesting environments that encompass the falls. As an image of Odisha's regular glory, Khandadhar turns into a wellspring of pride and association for the networks living in its area.

In Jharkhand, Lodh Falls divulges its ventured drop in the midst of the Chota Nagpur Level, offering a one of a kind component that recognizes it from different cascades. The falls, with its layered fountains, make a hypnotizing visual showcase as the waters explore through the rough scene. The ventured idea of Lodh adds to its appeal, introducing a dynamic and complex display against the background of Jharkhand's undulating territories.

The effect of Lodh on neighborhood networks reaches out to its job as a social and sporting space. The falls become a social event direct for networks toward celebrate conventional celebrations, cultivating a feeling of social congruity. Lodh, with its grand excellence, draws in guests and assumes a part in advancing the travel industry related exercises that benefit the neighborhood economy. The falls become a fundamental piece of the social character of Jharkhand, resounding with the tales and customs of the networks that call the level home.

In the northeastern territory of Tripura, Hattabari Cascade remains as a less popular pearl inside the verdant slopes close to Amarpur. The uniqueness of Hattabari lies in its detached appeal, stowed away from the standard vacationer circuits. The falls, with its peaceful fountain and the encompassing slopes, make a cozy and tranquil atmosphere that separates it from additional regularly visited cascades.

The effect of Hattabari on the nearby networks is entwined with its job as a characteristic retreat. The falls become a position of break for local people, offering a quiet background for thought and association with nature. Hattabari adds to the feeling of local area personality, with the falls filling in as an image of the pristine excellence that describes the scenes of Tripura. Neighborhood drives for the protection of Hattabari and its environmental elements turned into a demonstration of the job of the falls in cultivating ecological mindfulness.

Generally, these lesser-investigated cascades in the eastern piece of India have novel elements that stretch out past their actual excellence. Nyngnging in Arunachal Pradesh radiates quietness, offering a profound safe-haven in the midst of the Eastern Himalayas. Rangapahar in Nagaland remains as a segregated safe house, cultivating local meetings and ecological stewardship. Khandadhar in Odisha grandstands the crude force of nature, adding to far-reaching developments and monetary open doors. Lodh in Jharkhand presents a powerful scene, turning into a social and sporting point of convergence. Hattabari in Tripura encapsulates disconnected beguile, offering a peaceful retreat for neighborhood networks.

The effect of these cascades on neighborhood networks is multi-layered, going from social festivals and otherworldly importance to monetary open doors and natural preservation. As these fountains keep on shaping the lives and customs of the districts they effortlessness, they stand as demonstrations of the unpredictable connection among nature and culture, winding around a story that goes past the visual display of falling water to include the well established associations among networks and their normal legacy.

Proceeding with our investigation of the novel elements of lesser-investigated cascades in the eastern piece of India and their effect on neighborhood networks, we dive further into the rich embroidery of each fountain's impact. From the ancestral scenes of Arunachal Pradesh to the tranquil slopes of Tripura, these cascades not just add to the tasteful allure of the district yet in addition assume vital parts in forming the social, financial, and natural parts of the networks dwelling in their nearness.

In Arunachal Pradesh, Nyngnging Cascade's remarkable component of peacefulness reaches out past its actual presence. The falls become a consecrated space, entwined with the profound convictions of the native clans in the district. Neighborhood people group view Nyngnging as a wellspring of heavenly energy, and customs and services are led to look for endowments and offer thanks. The cascade turns into a living element, epitomizing the amicable concurrence among nature and individuals.

The effect of Nyngnging on the otherworldly texture of the networks is significant. It encourages a feeling of solidarity and respect, imparting an obligation to safeguard the normal sacredness of the falls. Social practices and customary services revolved around Nyngnging add to the protection old enough old convictions, passing down familial insight to people in the future.

The falls, with their profound importance, become an immortal demonstration of the social legacy of Arunachal Pradesh.

In Nagaland, Rangapahar Cascade's segregated magnificence holds the remarkable element of going about as a common social event space. Nearby people group use the falls as a scene for far-reaching developments, celebrations, and festivities. Rangapahar turns into a characteristic amphitheater where customs wake up, reverberating with the cadenced beats of native music and dance. The falls act as a scenery for narrating meetings, where stories of nearby legends and fables are shared, making a residing storehouse of Nagaland's social story.

The effect of Rangapahar on social safeguarding is obvious as it turns into a center point for passing down customs to more youthful ages. The falls encourage a feeling of local area pride, and drives for social supportability are frequently sent off. From the perspective of Rangapahar, Nagaland's people group track down a space to commend their personality, cultivating a social strength that endures everyday hardship.

Going to Odisha, Khandadhar Falls' novel element lies in its job as a magnet for the travel industry, adding to financial open doors for the nearby networks. The falls become an objective for nature devotees, experience searchers, and sightseers anxious to observe the glory of the Eastern Ghats. Nearby organizations, including cordiality, guides, and high quality specialties, flourish because of the deluge of guests attracted to Khandadhar.

The effect of Khandadhar on the neighborhood economy is significant. The falls animate work creation and business venture, engaging networks to reasonably exploit their normal assets. As the travel industry prospers, there is an equal work to safeguard the biological equilibrium encompassing Khandadhar. Nearby people group effectively participate in preservation projects, perceiving the reliance between the travel industry, the climate, and their jobs.

In Jharkhand, Lodh Falls' special ventured plunge makes an amazing visual scene. The falls become a material for creative articulation, drawing in photographic artists, painters, and nature fans. Nearby people group bridle the imaginative allure of Lodh to feature the social wealth of Jharkhand through different mediums. Craftsmanship celebrations and presentations praising the falls' magnificence add to a dynamic social scene.

The effect of Lodh on the creative and social scene of Jharkhand is critical. The falls move inventiveness and become an image of territorial character. Specialists, both neighborhood and visiting, track down motivation in the powerful presentation of flowing waters against the rough level. Lodh becomes a characteristic miracle as well as a dream for social articulation, encouraging a deep satisfaction and appreciation for human expressions inside the nearby networks.

In Tripura, Hattabari Cascade's novel isolated engage offers something other than a grand departure. The falls become a restorative space, advancing prosperity and psychological well-being for the neighborhood networks. The quiet mood and the thoughtful sound of falling water make a characteristic safe-haven where people look for comfort, reflection, and a brief departure from everyday burdens.

The effect of Hattabari on the psychological and close to home prosperity of the neighborhood networks is unlimited. The falls act as a retreat, offering snapshots of reprieve for those wrestling with the difficulties of current life. Neighborhood drives might arise to advance Hattabari as a wellbeing objective, where yoga withdraws, reflection meetings, and nature treatment add to the comprehensive strength of the local area. Hattabari turns into a demonstration of the characteristic association among nature and prosperity.

As we keep on disentangling the remarkable highlights and effects of these cascades, it becomes obvious that their importance goes past the simple visual exhibition. Nyngnging in Arunachal Pradesh turns into a consecrated material for profound customs and social safeguarding. Rangapahar in Nagaland changes into a social amphitheater, cultivating local meetings and narrating. Khandadhar in Odisha arises as a monetary impetus, driving the travel industry and occupation open doors. Lodh in Jharkhand turns into an imaginative dream, adding to the social dynamic quality of the area. Hattabari in Tripura rises above the picturesque, offering a restorative space for mental and profound prosperity.

These cascades are not inactive scenes but rather powerful elements that effectively shape the stories of the networks they occupy. They act as images of social character, financial potential, profound sacredness, creative motivation, and all encompassing prosperity. Fundamentally, the special highlights of these cascades entwine with the existences of the nearby networks, making a harmonious relationship that improves the social and ecological embroidery of the eastern piece of India.

4.3 Explore the interplay of culture, tradition, and nature in the East.

Investigating the exchange of culture, custom, and nature in the eastern piece of India reveals a rich embroidery where these components flawlessly entwine, making a dynamic and amicable relationship. From the verdant scenes of Arunachal Pradesh to the peaceful slopes of Tripura, the social practices and customs of the different networks in the East are well established in their regular environmental elements.

In Arunachal Pradesh, the Eastern Himalayas act as the background for a social scene that is essentially as different as its geography. The native clans of the district, including the Nyishi, Adi, Apatani, and Galo, have developed an extraordinary lifestyle complicatedly connected to the regular rhythms of the mountains, waterways, and woodlands that characterize their country.

The social acts of these clans are many times revolved around a significant regard for nature, mirroring a comprehension of the sensitive harmony between human life and the climate.

Customary celebrations in Arunachal Pradesh, like Losar, Solung, and Nyokum, are festivities that concur with the rural schedule and the evolving seasons. These celebrations include lively services, moves, and ceremonies that offer thanks to nature for its abundance and look for endowments for an agreeable concurrence. The mind boggling handloom and handiwork customs of the district, including the renowned ancestral weaves and bamboo creates, draw motivation from the normal materials plentifully tracked down in the Eastern Himalayas.

The transaction of culture, custom, and nature in Arunachal Pradesh is especially obvious in the profound acts of the neighborhood networks. Hallowed destinations, frequently settled in the core of immaculate scenes, become spots of respect and love. These destinations incorporate high mountain passes, old forests, and flowing cascades, like Nyngnging Cascade. The falls, with their tranquil outpouring, are an actual wonder as well as a profound safe-haven where customs and services associate individuals with the heavenly energy accepted to dwell in the regular environmental factors.

As we cross toward the east, Nagaland remains as a demonstration of the versatility of native societies in the midst of the slopes and valleys of the district. The Naga clans, with their particular customs and vernaculars, have developed areas of strength for a with nature that saturates each part of their day to day routines. The horticultural acts of the Naga people group are intently attached to the land, with terraced fields etched into the slopes mirroring a feasible way to deal with cultivating.

Social merriments in Nagaland, strikingly the Hornbill Celebration, exhibit the different customs of the Naga clans. The celebration, named after the Hornbill bird respected by the Nagas, is a lively festival of dance, music, and native specialties. It fills in as a stage for the safeguarding of customary works of art, exhibiting the multifaceted subtleties of Naga cloaks, gems, and woodcarvings, all motivated by the normal environmental factors of the district.

The transaction of culture, custom, and nature in Nagaland is well established in animistic convictions. Sacrosanct forests, antiquated trees, and normal arrangements hold profound importance, and functions directed in these spaces are articulations of appreciation and adoration for the climate. Rangapahar Cascade, concealed inside the slopes close to Dimapur, becomes a beautiful marvel as well as a social image, typifying the otherworldly associations of the Naga nation with their normal legacy.

Turning our concentration to Odisha, the state's social mosaic is formed by a combination of old customs and current impacts.

The Eastern Ghats, with their different environments, have been a support for civilizations that have prospered along the banks of streams like the Mahanadi and Godavari. Odisha's social legacy is obvious in its traditional dance structures, design, and customs that mirror a significant association with nature.

The dance types of Odissi, starting from the sanctuaries of the state, are a sign of the transaction among culture and otherworldliness. Moves frequently portray stories from Hindu folklore, commending the enormous dance of Ruler Shiva or the peaceful life related with the district. The Konark Sun Sanctuary, an UNESCO World Legacy site, isn't just a compositional wonder however a demonstration of the social blend of craftsmanship and nature, with its unpredictable carvings portraying scenes of day to day existence and fanciful stories.

Social celebrations in Odisha, like the Rath Yatra in Puri, unite a large number of fans to observe the excellent parade of chariots conveying divinities. The celebration, established in old customs, highlights the significance of regular components in the strict acts of the state. The sacrosanct Gundicha Sanctuary, where the divinities are taken during the Rath Yatra, turns into a brief house encompassed by the lavish scenes of Puri.

The exchange of culture, custom, and nature in Odisha is likewise obvious in the state's worship for its water bodies. The Chilika Lake, Asia's biggest bitter water tidal pond, isn't just a crucial biological system however a social center point. The lake upholds an interesting mix of fishing networks, birdlife, and conventional boat-production rehearses, all profoundly weaved with the rhythms of Chilika's waters.

Jharkhand, settled in the Chota Nagpur Level, epitomizes a social scene molded by its ancestral networks and the undulating territories of the district. The state's ancestral craftsmanship, including Dokra metalwork and Sohrai and Khovar painting, mirrors the harmonious connection among culture and nature. Dokra craftsmans draw motivation from the vegetation of the district, making mind boggling metal figures that catch the substance of Jharkhand's biodiversity.

Conventional celebrations in Jharkhand, for example, the Sarhul celebration celebrated by the Munda clan, are connected to horticultural cycles and the love of nature. Sarhul, set apart by the planting of seeds and the love of the hallowed tree, Sal, is a festival of fruitfulness and overflow. The celebration includes energetic music, dance, and ceremonies that feature the close association between ancestral networks and the land they occupy.

The exchange of culture, custom, and nature in Jharkhand stretches out to the state's normal marvels, including Lodh Falls. The ventured drop of Lodh, settled in the lavish vegetation of Netarhat, turns into a material for imaginative motivation and social imagery. The falls, encompassed by thick woods and rough scenes, bring out a feeling of stunningness and miracle that resounds with the ancestral networks of Jharkhand.

As we move further toward the east, Tripura unfurls as a state where social legacy is unpredictably woven into the slopes and valleys. The state's native networks, including the Tripuri and Reang clans, have safeguarded their practices in the midst of the green scenes. Tripura's social articulations find reverberation in its dance structures, music, and customs that draw motivation from the normal abundance of the district.

The Garia Puja, celebrated by the Tripuri people group, is a celebration devoted to the love of Master Garia, the divinity of riches and thriving. The ceremonies include the planting of Garia trees, which are accepted to be the residence of the god. The celebration mirrors the local area's well established association with the land and the agrarian cycles that support their lifestyle.

The transaction of culture, custom, and nature in Tripura is likewise clear in the state's native specialties. Bamboo, plentiful in the district, is unpredictably woven into customary ancient rarities, including basketry and material plans. The state's handloom customs, with energetic varieties and multifaceted examples, reverberation the normal liveliness of Tripura's scenes.

As we close our investigation of the exchange of culture, custom, and nature in the eastern piece of India, obviously these components are not disengaged yet profoundly interconnected. In Arunachal Pradesh, Nagaland, Odisha, Jharkhand, and Tripura, the regular environmental factors are not only backgrounds to social practices and customs; they are vital parts that shape the actual substance of life in these locales. The dance of culture, custom, and nature unfurls as an agreeable orchestra, where every component improves and supports the others, making a social scene that is both dynamic and persevering.

Proceeding with our investigation of the multifaceted exchange of culture, custom, and nature in the eastern piece of India, we dig into the nuanced articulations of these components in the district's assorted scenes, uncovering the manners by which networks weave their character with the normal woven artwork that encompasses them.

In Arunachal Pradesh, the blend of culture and nature stretches out to the building wonders that dab the scene. Conventional houses, for example, the unnatural bamboo structures normal for the Apatani clan, consistently mix with the verdant environmental elements. These homes, intended to endure the locale's climatic difficulties, become a demonstration of the cooperative connection between human residence and the indigenous habitat.

The exchange of culture and nature in Arunachal Pradesh is likewise apparent in the district's culinary practices. Native fixings obtained from the backwoods and waterways track down their direction into conventional dishes, making a culinary scene that isn't just flavorsome yet well established in the biological system. For example, bamboo shoot-based indulgences and native spices mirror a culinary legacy that reflects the rich biodiversity of the Eastern Himalayas.

As we adventure into Nagaland, the social texture of the state is unpredictably woven with the energetic materials created by Naga ladies. Customary wraps, enhanced with complicated examples and themes, frequently draw motivation from the vegetation of the area. The particular plans not just act as a type of creative articulation yet additionally as a visual language that conveys the Naga nation's association with nature.

In Nagaland, social practices, for example, inking hold profound importance, with designs roused by normal components like creatures and plants. These tattoos, inserted with social accounts, become living demonstrations of the entwined connection between the Naga clans and their current circumstance. The creative articulation through inking is a festival of nature's excellence and an approach to saving social personality.

Moving to Odisha, the state's social legacy is typified by the Pattachitra artistic expression. Unpredictably painted on fabric or palm leaves, Pattachitra frequently portrays scenes from Hindu folklore and day to day existence, integrating components of nature like trees, creatures, and waterways. The work of art turns into a visual story that rises above time, exhibiting the persevering through connection among culture and the normal world.

Odisha's customary music and dance structures, like Odissi, are profoundly affected by the state's social and regular legacy. Odissi, portrayed by its smooth motions and articulations, frequently draws motivation from the rhythms of streams, the influencing of trees, and the beauty of natural life. The dance turns into an idyllic articulation of the human association with the climate, rising above social limits.

In Jharkhand, the Dokra metalwork custom is a social sign well established in nature. Specialists, known as Dokras, make mind boggling metal models utilizing the lost-wax projecting method. The themes in Dokra workmanship frequently portray creatures, birds, and regular components, mirroring the native networks' veneration for the biodiversity of the Chota Nagpur Level.

Jharkhand's ancestral networks celebrate celebrations like Karma Puja, committed to the love of trees, mirroring a social beneficial interaction with the climate. The ceremonial dance during Karma Puja turns into a cadenced articulation of appreciation to nature for its overflow. The celebration represents how social customs in Jharkhand are indivisible from the normal world, blending with the patterns of horticulture and the evolving seasons.

As we venture further east, Tripura's social legacy tracks down articulation in the state's native artworks, especially bamboo and material expressions. The mind boggling winding around designs in Tripura's materials reflect the sensitive transaction of light and shadows in the timberlands, and bamboo creates, going from bushels to customary instruments, epitomize the genius of the networks in using normal materials.

The Garia Puja celebration in Tripura, committed to the divinity of abundance and flourishing, includes the love of hallowed trees, highlighting the social worship for nature. The ceremonies performed around these trees become a local area festivity, supporting the interconnectedness between culture, custom, and the climate.

In each condition of the eastern district, the conservation of social legacy is entwined with endeavors to preserve the common habitat. The idea of holy forests, present in Arunachal Pradesh and Tripura, mirrors a firmly established faith in the holiness of explicit forested regions. These forests, frequently home to different vegetation, act as residing stores of biodiversity and social works on, encouraging a comprehensive way to deal with preservation.

Local area based preservation drives, similar to those seen in Jharkhand, where customary celebrations are connected to ecological stewardship, show the way that social practices can turn out to be amazing assets for safeguarding regular assets. The respect for sacrosanct streams, for example, the Mahanadi in Odisha, further represents how social convictions can add to the safeguarding of crucial biological systems.

The exchange of culture, custom, and nature in the eastern piece of India is a dynamic and corresponding relationship. From the engineering of homes to the complicated plans in materials, from conventional expressions to consecrated ceremonies, the social articulations of the different networks in Arunachal Pradesh, Nagaland, Odisha, Jharkhand, and Tripura are well established in the normal scenes that encompass them.

The safeguarding of social legacy becomes inseparable from natural preservation, cultivating a feeling of obligation and stewardship toward the biological systems that support these networks. The amicable concurrence among culture and nature in the East isn't simply a verifiable heritage yet a living, developing story that keeps on forming the personality and strength of individuals in this socially different and environmentally rich district.

Chapter 5

Southern Rhythms
Waterfalls in the Deccan Plateau

Setting out on an excursion to investigate the cascades of the Deccan Level in southern India uncovers a hypnotizing ensemble of nature's greatness. The Deccan Level, a huge and raised district that traverses a few states, is embellished with flowing cascades, each having its own interesting appeal and character. From the Western Ghats toward the Eastern Ghats, this topographical wonder is accentuated by cascades that charm the faculties as well as assume necessary parts in forming the scene, nature, and social stories of the southern states.

In the core of the Deccan Level, settled in the midst of the lavish Western Ghats, lies Run Falls - a divine plunge of water that orders stunningness and love. Made by the Sharavathi Stream, Run Falls is a terrific outpouring with a level of more than 800 feet, making it perhaps of the tallest cascade in India. The falls, encompassed by immaculate timberlands and steep bluffs, present a visual display that rises above the common, summoning a feeling of marvel at the crude force of nature.

Run Falls, with its four unmistakable fountains - Raja, Rani, Roarer, and Rocket - becomes a characteristic milestone as well as a social symbol. The falls are unpredictably woven into the old stories and folklore of the district. Rumors from far and wide suggest that the divinities worshipped by the nearby networks dropped to favor the earth, and Run Falls represents their heavenly presence. Explorers and travelers the same are attracted to Run Falls, not exclusively to observe its stunning excellence yet additionally to participate in the profound atmosphere that pervades the environmental factors.

The effect of Run Falls on the neighborhood environment is significant. The perpetual progression of water shapes the geology, cutting through rocks and dregs to make the glorious dive pool at the foundation of the falls. The encompassing woods, supported by the fog and splash from Run Falls, harbor a rich biodiversity, making a safe house for endemic greenery. The falls, in their greatness, become an imperative part of the Western Ghats' fragile biological system, adding to the natural equilibrium of the district.

As we navigate toward the east, the Deccan Level uncovers one more gem in its crown - the ethereal Athirapally Falls in Kerala. Frequently alluded to as the "Niagara of India," Athirapally is a flowing wonder where the Chalakudy Stream smoothly dives from the Anamudi Mountains. Encircled by the verdant Sholayar Woods, Athirapally enamors with its amicable mix of water, rock, and plant life.

The special component of Athirapally lies in its job as a realistic background, having been highlighted in various Indian and global movies. The falls, with their realistic charm, have turned into a magnet for movie producers, adding to the social story related with this normal marvel. Athirapally's on-screen presence has, thusly, added to its prominence among sightseers, making it a sought-after objective for those looking for the wizardry of both nature and film.

The effect of Athirapally on the nearby networks goes past its visual allure. The falls act as a wellspring of occupation for the native clans in the district, who take part in eco-the travel industry drives and customary specialties. The encompassing woods, home to a different scope of widely varied vegetation, benefit from protection endeavors prodded by the financial open doors made by Athirapally. The falls hence become an impetus for supportable turn of events, blending the interests of neighborhood networks with the need to save the environmental uprightness of the area.

Wandering further into the Deccan Level, the captivating Dudhsagar Falls in Goa offers a display that consistently mixes regular excellence with social importance. Dudhsagar, meaning "Ocean of Milk," is a multi-layered cascade where the Mandovi Waterway falls over a precarious bluff in the midst of the Western Ghats. The falls get their name from the smooth foam made as the water dives, looking like a flowing white drape.

The exceptional element of Dudhsagar lies in its relationship with nearby legends and folklore. As per well known legend, the falls were shaped by the requests of an in the princess waters to purge herself. The flowing white froth is accepted to represent the immaculateness and holiness of her requests. This fanciful association adds a layer of social profundity to Dudhsagar, making it a characteristic marvel as well as a vault of stories and customs went down through ages.

Dudhsagar's effect on the social texture of Goa reaches out to its job in strict journeys. The falls are arranged in closeness to the Tambdi Surla Sanctuary, an old Shiva sanctuary encompassed by the Western Ghats. Explorers embrace journeys to Dudhsagar as a feature of strict ceremonies, making an otherworldly association between the regular and the heavenly. The falls, encompassed by the lavish plant life of the Bhagwan Mahavir Untamed life Safe-haven, become a journey site that rises above the limits of ordinary strict practices.

The biological effect of Dudhsagar is important, adding to the biodiversity of the Western Ghats. The encompassing timberlands act as living spaces for different natural life, including endemic species. The Mandovi Stream, which brings forth Dudhsagar, likewise assumes an essential part in the hydrology of the locale. The falls, in their grand plunge, become fundamental to the biological equilibrium of the Deccan Level.

Moving toward the east, the Deccan Level uncovers the shocking Sivasamudram Falls in Karnataka, where the Kaveri Stream goes all in. Sivasamudram is a fragmented cascade, comprising of two fundamental fountains - Barachukki and Gaganachukki. The falls, encompassed by rough territory and thick vegetation, epitomize the unique power of water as it flows through the Deccan scene.

The exceptional element of Sivasamudram lies in its authentic importance as one of the earliest hydroelectric power stations in Asia. Worked during the English pioneer time in the mid twentieth hundred years, the falls assumed a urgent part in saddling the force of water to produce power. The designing wonder, joined with the normal magnificence of the falls, adds a layer of social and verifiable significance to Sivasamudram.

Sivasamudram's effect on the locale's social story reaches out to its job in water system. The Kaveri Waterway, which takes care of the falls, is a life saver for horticulture in the encompassing regions. The water from Sivasamudram is diverted through an organization of trenches to flood immense stretches of farmland, adding to the agrarian economy of Karnataka. The falls, while representing the untamed magnificence of nature, additionally epitomize the advantageous connection between water assets and human progress.

As we dive further into the Deccan Level, the Kiliyur Falls in Tamil Nadu arises as an unlikely treasure concealed in the Shevaroy Slopes. Kiliyur, portrayed by its thin drop over a rough incline, offers a quiet and detached climate. Encircled by thick timberlands, the falls make a peaceful retreat for those looking for reprieve from the hurrying around of metropolitan life.

The special component of Kiliyur lies in job as a characteristic safe-haven encourages biodiversity. The falls, settled inside the Yercaud Untamed life Asylum, add to the preservation of endemic widely varied vegetation. The encompassing slopes, covered with lavish vegetation, become a living space for various plant and creature species. Kiliyur Falls, with its unblemished environmental factors, becomes a visual joy as well as a demonstration of the significance of saving normal safe houses inside the Deccan Level.

Kiliyur's effect on the neighborhood networks is reflected in its job as a sporting and eco-the travel industry objective. The falls draw in nature fans, travelers, and experience searchers, giving a financial lift to the district. Nearby drives for mindful the travel industry and preservation endeavors add to the feasible advancement of the Shevaroy Slopes. Kiliyur Falls, with its fragile magnificence, turns into a guide for fitting the travel industry with natural stewardship.

Proceeding with our investigation, the Deccan Level unfurls the superb Hogenakkal Falls in Tamil Nadu, frequently alluded to as the "Niagara of the East." Cut by the Kaveri Waterway, Hogenakkal is a progression of fountains and rapids encompassed by rough bluffs and rich vegetation. The falls, with their wild plummet, make a scene that is both remarkable and helpful.

The extraordinary element of Hogenakkal lies in its relationship with conventional restorative practices. The waters of the falls are accepted to have recuperating properties, and the neighborhood networks have been involving them for quite a long time for helpful purposes. The presence of different minerals in the water is remembered to add to its restorative viability. Hogenakkal, subsequently, becomes a characteristic miracle as well as a health objective with social roots profoundly implanted in customary mending rehearses.

Hogenakkal's effect on the neighborhood economy is entwined with its job in medical care. The falls draw in guests looking for elective and comprehensive treatments, cultivating the development of wellbeing the travel industry in the locale. Neighborhood people group, frequently talented in customary medication, become necessary to the medical services contributions around Hogenakkal. The falls, encompassed by home grown gardens and native greenery, make a comprehensive encounter that orchestrates the mending force of nature with social wellbeing rehearses.

As we finish up our investigation of the cascades in the Deccan Level, it becomes apparent that these fountains are topographical arrangements as well as essential parts of the locale's social, environmental, and financial texture. Run Falls in Karnataka remains as a demonstration of the otherworldly and legendary importance entwined with its superb plunge. Athirapally Falls in Kerala turns into a true to life muse, winding around a social story from the perspective of film. Dudhsagar Falls in Goa unfurls as a social and strict image, mixing fantasy with regular excellence.

Sivasamudram Falls in Karnataka addresses a verifiable milestone, displaying the combination of designing resourcefulness with nature's power. Kiliyur Falls in Tamil Nadu remains as a segregated safe house, cultivating local meetings and natural stewardship. Hogenakkal Falls in Tamil Nadu exemplifies remedial appeal, offering a quiet retreat for nearby networks.

The effect of these cascades on neighborhood networks is complex, going from social festivals and otherworldly importance to monetary open doors and ecological protection. As these fountains keep on profoundly shaping the lives and customs of the locales they elegance, they stand as demonstrations of the complicated connection among nature and culture, winding around a story that goes past the visual display of falling water to include the well established associations among networks and their regular legacy. The Deccan Level, with its variety of cascades, becomes a land wonder as well as a material whereupon the narratives of southern India unfurl, mixing the domains of nature and culture into an agreeable embroidery.

5.1 Investigate the waterfalls that grace the Deccan Plateau in southern India.

Setting out on an analytical excursion into the cascades that elegance the Deccan Level in southern India reveals a spellbinding story of geographical marvels, social lavishness, and environmental importance. The Deccan Level, a broad raised locale traversing a few states, brags a different cluster cascades that entrance with their flowing excellence as well as add to the embroidery of the southern scene. From the Western Ghats toward the Eastern Ghats, every cascade recounts a special story of its development, social importance, and environmental effect, making the Deccan Level an intriguing material of regular wonders.

Run Falls, settled in the core of Karnataka in the midst of the Western Ghats, arises as a heavenly outpouring that orders consideration with its grand drop. Framed by the Sharavathi Stream, Run Falls remains as perhaps of the tallest cascade in India, with a level surpassing 800 feet. The falls divulge a terrific showcase of four unmistakable fountains - Raja, Rani, Roarer, and Rocket - by and large framing a stunning scene of nature's loftiness.

Run Falls, past being a geographical wonder, is complicatedly woven into the social texture of Karnataka. The falls track down reverberation in neighborhood old stories and folklore, where it is accepted that Run Falls addresses the heavenly plummet of gods who favored the land. Travelers and sightseers the same are attracted to Run Falls, not just to observe the crude force of water yet additionally to participate in the otherworldly atmosphere that wraps the environmental elements.

The social effect of Run Falls reaches out to its job in neighborhood festivities and celebrations. The falls become a focal point during strict parades and customary occasions, encapsulating a consecrated association between individuals and the normal world.

The environmental meaning of Run Falls is similarly significant, molding the scene and encouraging biodiversity in the Western Ghats. The perpetual progression of water shapes the geology, making a dive pool and supporting the encompassing woods, which, thus, harbor a rich exhibit of vegetation endemic to the locale.

Changing toward the east, Athirapally Falls in Kerala arises as a flowing magnum opus settled in the hug of the Western Ghats. Frequently named the "Niagara of India," Athirapally charms with its agreeable mix of water, rock, and plant life. Framed by the Chalakudy Stream, the falls slip smoothly from the Anamudi Mountains, making an ethereal display that deserves it acknowledgment as quite possibly of the most pleasant cascade in the country.

The novel element of Athirapally stretches out past its visual enticement for its job as a realistic scenery. The falls play had a conspicuous impact in various Indian and global movies, adding a layer of social story to its normal charm. The on-screen presence of Athirapally has, thus, added to its prominence among travelers, making it a sought-after objective for those looking for the enchantment of both nature and film.

The social and financial effect of Athirapally on neighborhood networks is obvious in its job as a wellspring of job. Native clans in the area take part in eco-the travel industry drives, offering directed visits and customary artworks to guests. The falls, encompassed by the Sholayar Timberland, become a characteristic marvel as well as a social and monetary impetus for the networks living in its closeness.

Further inland, Dudhsagar Falls in Goa reveals a staggering fountain that flawlessly mixes regular magnificence with social importance. Dudhsagar, meaning "Ocean of Milk," is a multi-layered cascade where the Mandovi Stream plunges over a precarious precipice in the midst of the Western Ghats. The falls get their name from the smooth foam made as the water slips, looking like a flowing white drape.

The one of a kind element of Dudhsagar lies in its relationship with nearby fables and folklore. Rumors have spread far and wide suggesting that the falls were shaped by the requests of an in the princess waters to purge herself. The flowing white froth is accepted to represent the immaculateness and sacredness of her requests. This fanciful association adds a layer of social profundity to Dudhsagar, making it a characteristic marvel as well as a storehouse of stories and customs went down through ages.

Dudhsagar's social effect stretches out to its part in strict journeys. The falls are arranged in nearness to the Tambdi Surla Sanctuary, an old Shiva sanctuary encompassed by the Western Ghats. Pioneers embrace journeys to Dudhsagar as a component of strict ceremonies, making an otherworldly association between the normal and the heavenly. The falls, wrapped by the rich vegetation of the Bhagwan Mahavir Untamed life Safe-haven, become a journey site that rises above the limits of regular strict practices.

The natural effect of Dudhsagar is imperative, adding to the biodiversity of the Western Ghats. The encompassing woodlands act as natural surroundings for different untamed life, including endemic species. The Mandovi Waterway, which brings forth Dudhsagar, likewise assumes a urgent part in the hydrology of the locale. The falls, in their superb plunge, become fundamental to the environmental equilibrium of the Deccan Level.

Wandering into Karnataka, Sivasamudram Falls unfurls as a powerful demonstration of the power of nature interweaved with verifiable importance. Arranged on the Kaveri Waterway, Sivasamudram is a portioned cascade comprising of two principal overflows - Barachukki and Gaganachukki. The falls, encompassed by rough territory and thick vegetation, represent the powerful power of water as it flows through the Deccan scene.

The one of a kind element of Sivasamudram lies in its verifiable significance as one of the earliest hydroelectric power stations in Asia. Worked during the English pilgrim time in the mid twentieth 100 years, the falls assumed a vital part in tackling the force of water to produce power. The designing wonder, joined with the regular excellence of the falls, adds a layer of social and verifiable significance to Sivasamudram.

Sivasamudram's social effect is obvious in its job as an image of progress and advancement. The falls, while holding their regular wonder, address the combination of mechanical development with the powers of nature. The encompassing scenes, molded by the progression of the Kaveri Waterway, become a demonstration of the amicable conjunction of human resourcefulness and the common habitat.

The falls' effect on the nearby economy reaches out to its job in water system. The Kaveri Stream, which takes care of the falls, is a help for horticulture in the encompassing regions. The water from Sivasamudram is directed through an organization of trenches to flood tremendous stretches of farmland, adding to the agrarian economy of Karnataka. The falls, while representing the untamed magnificence of nature, additionally epitomize the harmonious connection between water assets and human civilization.

Diving into Tamil Nadu, Kiliyur Falls arises as an unlikely treasure concealed in the Shevaroy Slopes, offering a peaceful and separated feeling. Portrayed by its slim drop over a rough cliff, Kiliyur makes a serene retreat for those looking for rest from the buzzing about of metropolitan life. Encircled by thick woodlands, the falls give an unblemished shelter inside the Yercaud Untamed life Safe-haven.

The one of a kind element of Kiliyur lies in job as a characteristic safe-haven encourages biodiversity. The falls, settled inside the asylum, add to the protection of endemic vegetation. The encompassing slopes, covered with rich vegetation, become a natural surroundings for different plant and creature species.

Kiliyur Falls, with its immaculate environmental elements, becomes a visual pleasure as well as a demonstration of the significance of saving regular sanctuaries inside the Deccan Level.

Kiliyur's effect on the nearby networks is reflected in its job as a sporting and eco-the travel industry objective. The falls draw in nature fans, travelers, and experience searchers, giving a financial lift to the area. Nearby drives for dependable the travel industry and protection endeavors add to the reasonable advancement of the Shevaroy Slopes. Kiliyur Falls, with its fragile excellence, turns into a guide for fitting the travel industry with natural stewardship.

Proceeding with our investigation, Hogenakkal Falls in Tamil Nadu arises as a glorious series of fountains and rapids, frequently alluded to as the "Niagara of the East." Cut by the Kaveri Stream, Hogenakkal makes an exhibition that is both sensational and restorative. The falls, encompassed by rough precipices and rich vegetation, offer a visual and tactile experience that enraptures guests.

The special element of Hogenakkal lies in its relationship with customary restorative practices. The waters of the falls are accepted to have mending properties, and the neighborhood networks have been involving them for quite a long time for helpful purposes. The presence of different minerals in the water is remembered to add to its restorative adequacy. Hogenakkal, thusly, becomes a characteristic marvel as well as a health objective with social roots profoundly implanted in customary mending rehearses.

Hogenakkal's effect on the nearby economy is entwined with its part in medical care. The falls draw in guests looking for elective and all encompassing treatments, encouraging the development of wellbeing the travel industry in the district. Nearby people group, frequently gifted in customary medication, become fundamental to the medical services contributions around Hogenakkal. The falls, encompassed by home grown gardens and native verdure, make a comprehensive encounter that blends the recuperating force of nature with social wellbeing rehearses.

As we close our examination concerning the cascades gracing the Deccan Level in southern India, it becomes clear that each outpouring isn't simply a geographical peculiarity yet a mind boggling transaction of social, natural, and monetary elements. Run Falls in Karnataka remains as a demonstration of the profound and legendary importance entwined with its grand plummet. Athirapally Falls in Kerala turns into a realistic dream, winding around a social story from the perspective of film. Dudhsagar Falls in Goa unfurls as a social and strict image, mixing fantasy with regular magnificence. Sivasamudram Falls in Karnataka addresses a verifiable milestone, exhibiting the combination of designing creativity with nature's power. Kiliyur Falls in Tamil Nadu remains as a confined shelter, cultivating local meetings and natural stewardship. Hogenakkal Falls in Tamil Nadu embodies helpful appeal, offering a quiet retreat for nearby networks.

The effect of these cascades on neighborhood networks is multi-layered, going from social festivals and profound importance to monetary open doors and natural preservation. As these fountains keep on shaping the lives and customs of the locales they beauty, they stand as demonstrations of the perplexing connection among nature and culture, winding around a story that goes past the visual display of falling water to envelop the well established associations among networks and their normal legacy. The Deccan Level, with its variety of cascades, becomes a geographical wonder as well as a material whereupon the tales of southern India unfurl, mixing the domains of nature and culture into an amicable embroidery.

Proceeding with our examination concerning the cascades that effortlessness the Deccan Level in southern India, we dig into the novel elements and social subtleties of each fountain, disentangling the assorted accounts that shape the area's personality.

Run Falls, with its transcending presence in Karnataka, rises above its land grandness to turn into a social symbol profoundly imbued in nearby practices. The falls' four unmistakable fountains - Raja, Rani, Roarer, and Rocket are not simply an exhibition for spectators but rather hold significant profound importance. As per nearby fables, Run Falls is accepted to be the natural sign of divinities sliding to favor the land. This fanciful association changes the falls into a sacrosanct site, drawing explorers looking for profound comfort.

The social effect of Run Falls reaches out to the dynamic festivals that unfurl in its area. The falls become a point of convergence during strict parades and customary occasions, representing an agreeable connection between the regular and otherworldly domains. Explorers attempt excursions to Run Falls, to observe its stunning magnificence as well as to participate in the heavenly energy that exudes from the flowing waters. Hence, the falls become a living demonstration of the social combination of nature and otherworldliness in the Deccan Level.

Athirapally Falls in Kerala, frequently alluded to as the "Niagara of India," presents an alternate feature of social commitment. Past its grand charm, Athirapally has turned into a dream for producers, adding to its ubiquity among film fans. The falls' realistic appearances in different Indian and worldwide movies have raised its social importance, transforming it into a famous setting that resounds with both neighborhood and worldwide crowds.

The social effect of Athirapally stretches out to the financial open doors it makes for nearby networks. Native clans in the district take part in eco-the travel industry drives, offering directed visits that give experiences into the rich biodiversity of the Sholayar Timberland encompassing the falls. Furthermore, conventional specialties and high quality items find a market among travelers looking for legitimate social encounters. Athirapally, thusly, becomes a characteristic marvel as well as a social and monetary impetus, supporting practical jobs.

Dudhsagar Falls in Goa, with its legendary history and strict meanings, remains as a demonstration of the social entwining of fantasy and nature. The falls, named "Ocean of Milk" for the smooth foam made as the water plunges, are related with a princess' requests for immaculateness. This fanciful association adds a layer of social profundity, transforming Dudhsagar into a storehouse of stories and customs went down through ages.

Dudhsagar's social effect reaches out to its part in strict journeys, making an agreeable combination of nature and otherworldliness. Travelers journey to the falls as a feature of customs related with the Tambdi Surla Sanctuary, making an other-worldly association between the regular and the heavenly. The falls, encompassed by the Bhagwan Mahavir Untamed life Safe-haven, become a journey site that rises above the limits of traditional strict works on, mixing fantasy, religion, and normal excellence.

Sivasamudram Falls in Karnataka, past its stunning excellence, unfurls as a verifiable milestone that represents the combination of designing creativity with nature's power. As one of the earliest hydroelectric power stations in Asia, the falls hold social importance by addressing progress and advancement. The falls, while holding their regular quality, become a demonstration of the agreeable concurrence of human development and the crude force of water.

The social effect of Sivasamudram is obvious in its portrayal as an image of mechanical progression. The falls draw in nature fans as well as those captivated by the authentic and designing parts of the locale. Guests draw in with the double story of nature's magnificence and human accomplishment, encouraging an all encompassing appreciation for the convergence of culture and innovation.

Kiliyur Falls in Tamil Nadu, settled inside the Shevaroy Slopes, presents a segregated shelter that encourages local meetings and ecological stewardship. Its slim plummet over a rough incline makes a serene retreat for those looking for reprieve from metropolitan life. Kiliyur Falls, encompassed by thick woodlands inside the Yercaud Untamed life Safe-haven, turns into a demonstration of the significance of protecting normal sanctuaries inside the Deccan Level.

The novel component of Kiliyur lies in job as a characteristic safe-haven encourages biodiversity. The falls, settled inside the asylum, add to the preservation of endemic verdure. Nearby drives for dependable the travel industry and protection endeavors make a harmonious connection between the falls and the networks. Kiliyur turns into a reference point for orchestrating the travel industry with ecological stewardship, featuring the social benefit of saving unblemished scenes.

Hogenakkal Falls in Tamil Nadu, frequently named the "Niagara of the East," divulges helpful fascinate that entwines with customary restorative practices.

The falls are accepted to have mending properties, and neighborhood networks have used their waters for a really long time for restorative purposes. Hogenakkal, consequently, becomes a characteristic miracle as well as a health objective with social roots profoundly implanted in customary mending rehearses.

The social effect of Hogenakkal stretches out to its part in medical care and wellbeing the travel industry. The falls draw in guests looking for elective and all encompassing treatments, cultivating the development of wellbeing related exercises in the district. The encompassing natural nurseries and native verdure add layers of social importance to the helpful experience. Hogenakkal turns into a space where nature's recuperating power converges with social wellbeing works on, making a comprehensive objective.

As we research these cascades gracing the Deccan Level, their remarkable highlights and social subtleties exhibit the unpredictable connection among nature and human social orders. Run Falls, Athirapally Falls, Dudhsagar Falls, Sivasamudram Falls, Kiliyur Falls, and Hogenakkal Falls become topographical miracles as well as social touchpoints that characterize the personality of southern India. Each fountain, with its own folklore, imagery, and monetary job, adds to the rich woven artwork of the Deccan Level, winding around together accounts of otherworldliness, custom, progress, and ecological stewardship. The examination concerning these cascades disentangles a story where the normal and social domains are indistinguishable, making a significant feeling of spot and legacy inside the flowing scenes of southern India.

5.2 Discuss the geological formations and historical significance of these cascades.

Setting out on an extensive investigation of the cascades that enhance the Deccan Level in southern India involves unwinding the geographical marvels as well as the significant verifiable importance implanted inside each fountain. This excursion takes us through an embroidery of scenes formed by time, normal powers, and the social stories that have developed around these superb falls.

Run Falls in Karnataka, remaining as perhaps of the tallest cascade in India, is a demonstration of the topographical powers that have shaped the Western Ghats. The falls owe their reality to the Sharavathi Waterway, which, over ages, has cut through the rough landscape, making a cliff from which the waters plunge emphatically. The geographical arrangement of Run Falls is portrayed by the presence of hard rock developments, basically made out of stone, through which the waterway has scratched its way.

The one of a kind topographical component of Run Falls lies in its fragmented drop, containing four unmistakable fountains - Raja, Rani, Roarer, and Rocket.

Every one of these falls adds to the general exhibition, making an orchestra of water that reverberates with the complex land cosmetics of the locale. The tireless disintegration of the gentler stone layers by the Sharavathi Stream plays had a urgent impact in forming the personality of Run Falls, where the water emphatically jumps over the precipices, making a permanent imprint on the scene.

Past its topographical heavenliness, Run Falls has a rich verifiable embroidery entwined with the social texture of Karnataka. The falls track down notice in antiquated texts, and neighborhood fables describes enrapturing accounts of divinities dropping to favor the land. The verifiable meaning of Run Falls rises above simple perception, welcoming pioneers and explorers to participate in the otherworldly energy accepted to exude from the flowing waters.

Athirapally Falls in Kerala, frequently alluded to as the "Niagara of India," uncovers a geographical story formed by the Chalakudy Stream and the Anamudi Mountains. The falls owe their starting point to the rough shapes of the Western Ghats, where the stream arranges its direction through the undulating territory, making a fountain that spellbinds with its amicable mix of water and rock. The land arrangement of Athirapally is set apart by the lofty inclinations of the Western Ghats, permitting the Chalakudy Waterway to flow over the cliff with sheer power.

The stones encompassing Athirapally Falls bear the engravings of time, showing the topographical history of the district. The falls, with their sectioned plunge and the wild progression of water, mirror the unique geographical cycles that have formed the scene over centuries. The presence of basaltic rocks and the exchange of geographical powers add to the visual show of Athirapally, making it a land magnum opus inside the Western Ghats.

The authentic meaning of Athirapally is entwined with the old developments that flourished in the area. The falls and their environmental elements have been observer to the back and forth movement of societies, with native clans tracking down food and motivation from the bounties of nature. The geographical developments encompassing Athirapally have formed the actual scene as well as impacted the social stories that have advanced over hundreds of years.

Dudhsagar Falls in Goa, getting its name from the smooth foam looking like a "Ocean of Milk," discloses a topographical scene etched by the Mandovi Stream. The falls overflow over a precarious bluff in the midst of the Western Ghats, making a visual exhibition that mirrors the topographical variety of the locale. The topographical development of Dudhsagar is portrayed by the connection between the stream and the rough territory, where the waters plunge decisively, making a hypnotizing show of nature's powers.

The stones encompassing Dudhsagar bear the scars of topographical development, with the erosive force of the Mandovi Waterway scratching its way through the scene. The falls, with their multi-layered drop, epitomize the geographical intricacy of the Western Ghats, where the territory impacts how water navigates the scene.

The land arrangements of Dudhsagar contribute not exclusively to its visual loftiness yet additionally to the natural meaning of the Bhagwan Mahavir Untamed life Asylum that wraps the falls.

The authentic meaning of Dudhsagar is well established in nearby legends and folklore. As per legend, the falls were framed by the requests of an in the princess waters to scrub herself. The flowing white froth represents the immaculateness and sacredness of her requests, adding a layer of verifiable and social profundity to Dudhsagar. The falls, encompassed by the rich plant life of the Bhagwan Mahavir Natural life Safe-haven, become a storehouse of stories and customs went down through ages.

Sivasamudram Falls in Karnataka, arranged on the Kaveri Stream, unfurls as a geographical wonder formed by the many-sided dance among water and rock. The falls are portioned into two fundamental fountains - Barachukki and Gaganachukki - each adding to the visual exhibition that portrays Sivasamudram. The topographical development of the falls is impacted by the translucent rocks that structure the bedrock of the Deccan Level, making a material whereupon the Kaveri Waterway paints its flowing drop.

The stones encompassing Sivasamudram take the stand concerning the erosive powers of the Kaveri, with the waterway chiseling its direction through the scene over centuries. The falls, with their booming plummet, epitomize the geographical history of the Deccan Level, where the cooperation among water and rock has brought about amazing scenes. The land highlights of Sivasamudram contribute not exclusively to its visual appeal yet in addition to the verifiable meaning of the district.

The verifiable significance of Sivasamudram is set apart by its job as one of the earliest hydroelectric power stations in Asia. Worked during the English frontier time in the mid twentieth 100 years, the falls turned into a spearheading site tackling the force of water for power age. The designing wonder, joined with the normal magnificence of the falls, adds a layer of verifiable and social significance to Sivasamudram, exhibiting the crossing point of mechanical advancement with the powers of nature.

Kiliyur Falls in Tamil Nadu, settled inside the Shevaroy Slopes, divulges a land show-stopper that mirrors the district's geological complexities. The falls, described by their thin plummet over a rough slope, encapsulate the topographical variety of the Deccan Level. The stones encompassing Kiliyur bear the characteristics of enduring and disintegration, giving bits of knowledge into the geographical cycles that have molded the scene over the long run.

The topographical development of Kiliyur is affected by the rough outcrops inside the Yercaud Natural life Asylum. The falls, with their fragile drop, grandstand the delicate yet industrious powers that shape the rough landscape.

The geographical elements of Kiliyur contribute not exclusively to its tasteful appeal yet additionally to the natural meaning of the encompassing slopes, filling in as an environment for different verdure.

The verifiable meaning of Kiliyur is woven into the social stories of the district. The falls, disconnected inside the safe-haven, have been an observer to the recurring pattern of developments that have looked for shelter in the Shevaroy Slopes. The verifiable layers encompassing Kiliyur add profundity to its social significance, as it keeps on being a position of comfort and examination for those looking for the peacefulness of nature.

Hogenakkal Falls in Tamil Nadu, frequently named the "Niagara of the East," arises as a topographical miracle molded by the tireless progression of the Kaveri Waterway. The falls, encompassed by rough precipices and rich vegetation, embody the topographical variety of the Deccan Level. The stones encompassing Hogenakkal bear the engravings of water's erosive power, making a scene where nature's masterfulness is in plain view.

The geographical development of Hogenakkal is portrayed by the stream's dive over the rough landscape, making a progression of fountains and rapids. The remarkable geology of the area, with its rough outcrops and undulating scene, impacts how the falls unfurl. The geographical elements of Hogenakkal contribute not exclusively to its visual magnificence yet additionally to the remedial enchant related with its waters.

Digging into the verifiable meaning of Hogenakkal divulges a social story well established in customary restorative practices. The falls are accepted to have recuperating properties, and neighborhood networks have been bridling their restorative potential for a really long time. The verifiable layers encompassing Hogenakkal add social profundity to its character, as the falls keep on being a space where nature's mending power converges with conventional health rehearses.

As we investigate these cascades gracing the Deccan Level, their land developments arise as powerful materials formed by the perplexing transaction of water and rock. Run Falls, Athirapally Falls, Dudhsagar Falls, Sivasamudram Falls, Kiliyur

Falls, and Hogenakkal Falls stand as topographical wonders that enthrall with their visual quality as well as recount the account of the World's extraordinary powers over the long run.

The verifiable meaning of these cascades, profoundly imbued in neighborhood fables, folklore, and human undertakings, adds layers of social wealth to their personality. Run Falls, with its divine plunge, turns into a profound safe-haven where the verifiable association among divinities and the land is commended. Athirapally Falls, with its true to life charm, turns into a social material where the visual expressions interweave with the land material of the Western Ghats.

Dudhsagar Falls, with its fanciful roots and strict meanings, turns into a verifiable storehouse where the requests of a princess reverberation through time. Sivasamudram Falls, with its spearheading job in hydroelectric power age, turns into a verifiable milestone where innovation merges with the basic powers of nature. Kiliyur Falls, confined inside the Shevaroy Slopes, turns into a verifiable observer to the ages of normal cycles molding its land character.

Hogenakkal Falls, with its remedial appeal and customary mending rehearses, turns into a verifiable safe-haven where the restorative insight of neighborhood networks interweaves with the geographical legacy of the Deccan Level. Every one of these cascades, in its land greatness and verifiable reverberation, turns into a part in the excellent story of the Deccan Level, where the World's powers and human stories meet in a hypnotizing dance of time and nature.

Proceeding with our profound plunge into the geographical developments and authentic meaning of the cascades gracing the Deccan Level, we currently direct our concentration toward the excess fountains - Sivasamudram Falls, Kiliyur Falls, and Hogenakkal Falls - disentangling the layers that make every one of them one of a kind inside the complicated embroidery of southern India.

Sivasamudram Falls, situated on the Kaveri Waterway in Karnataka, isn't just a geographical wonder yet in addition a verifiable milestone with a spearheading job in hydroelectric power age. The falls are divided into two fundamental fountains - Barachukki and Gaganachukki - each adding to the visual display that portrays Sivasamudram. The topographical development of the falls is impacted by the translucent rocks that structure the bedrock of the Deccan Level, making a material whereupon the Kaveri Waterway paints its flowing plunge.

The stones encompassing Sivasamudram give testimony regarding the erosive powers of the Kaveri, with the waterway chiseling its direction through the scene over centuries. The falls, with their loud plummet, epitomize the geographical history of the Deccan Level, where the communication among water and rock has brought about stunning scenes. The geographical highlights of Sivasamudram contribute not exclusively to its visual appeal yet additionally to the verifiable meaning of the district.

Sivasamudram's authentic significance is set apart by its change into one of the earliest hydroelectric power stations in Asia during the English frontier time. The development of force plants in the mid twentieth century saddled the energy of the flowing waters, spearheading the use of water assets for power age. This convergence of mechanical advancement and normal powers adds a layer of verifiable significance to Sivasamudram, displaying the locale's commitment to the development of energy creation.

The falls, while holding their normal wonder, turned into an image of innovation and progress. The designing wonder addressed an amicable concurrence between nature's powers and human inventiveness.

Today, Sivasamudram Falls stands as a geographical show-stopper as well as a verifiable demonstration of the extraordinary effect of human undertakings on the scene.

Kiliyur Falls in Tamil Nadu, settled inside the Shevaroy Slopes, presents a novel land development that mirrors the district's geological complexities. The falls, described by their thin drop over a rough incline, epitomize the geographical variety of the Deccan Level. The stones encompassing Kiliyur bear the characteristics of enduring and disintegration, giving bits of knowledge into the topographical cycles that have molded the scene over the long haul.

The topographical arrangement of Kiliyur is impacted by the rough outcrops inside the Yercaud Natural life Asylum. The falls, with their sensitive drop, exhibit the delicate yet industrious powers that shape the rough landscape. The land elements of Kiliyur contribute not exclusively to its stylish appeal yet in addition to the biological meaning of the encompassing slopes, filling in as a living space for different greenery.

The authentic meaning of Kiliyur is woven into the social stories of the locale. The falls, detached inside the safe-haven, have been an observer to the recurring pattern of civilizations that have looked for shelter in the Shevaroy Slopes. The verifiable layers encompassing Kiliyur add profundity to its social significance, as it keeps on being a position of comfort and thought for those looking for the serenity of nature.

Kiliyur Falls likewise assumes a critical part in neighborhood eco-the travel industry drives. The falls draw in nature lovers, travelers, and experience searchers, adding to the monetary development of the area. Nearby people group participate in dependable the travel industry works on, guaranteeing the conservation of the sensitive biological equilibrium encompassing the falls. Kiliyur, with its land and verifiable importance, becomes a characteristic miracle as well as an impetus for supportable the travel industry and natural stewardship.

Hogenakkal Falls in Tamil Nadu, frequently named the "Niagara of the East," arises as a topographical miracle molded by the determined progression of the Kaveri Stream. The falls, encompassed by rough bluffs and rich vegetation, encapsulate the geographical variety of the Deccan Level. The stones encompassing Hogenakkal bear the engravings of water's erosive power, making a scene where nature's creativity is in plain view.

The geographical development of Hogenakkal is described by the stream's dive over the rough territory, making a progression of fountains and rapids. The novel geography of the district, with its rough outcrops and undulating scene, impacts how the falls unfurl. The geographical highlights of Hogenakkal contribute not exclusively to its visual greatness yet in addition to the helpful enchant related with its waters.

The restorative properties of Hogenakkal's waters, accepted to have recuperating characteristics, add a remarkable social layer to the falls. Neighborhood people group, frequently talented in customary medication, use the falls as a space for all encompassing recuperating rehearses. The presence of different minerals in the water is remembered to add to its restorative viability. Hogenakkal, thusly, becomes a topographical marvel as well as a health objective with social roots profoundly implanted in conventional mending rehearses.

Hogenakkal Falls likewise assumes a critical part in the nearby economy through health the travel industry. The falls draw in guests looking for elective and comprehensive treatments, cultivating the development of wellbeing related exercises in the district. Neighborhood people group, with their insight into conventional medication, become basic to the medical care contributions around Hogenakkal. The falls, encompassed by home grown gardens and native vegetation, make a comprehensive encounter that orchestrates the mending force of nature with social wellbeing rehearses.

As we close our investigation of Sivasamudram Falls, Kiliyur Falls, and Hogenakkal Falls, the topographical developments and verifiable meaning of these fountains arise as spellbinding accounts inside the bigger story of the Deccan Level. Every cascade, with its remarkable land highlights and social subtleties, adds to the rich embroidery of southern India, displaying the significant exchange between normal powers and mankind's set of experiences. From hydroelectric power stations to helpful safe-havens, these falls epitomize the complex associations between geographical marvels and the social character of the districts they elegance.

5.3 Showcase the diversity of landscapes and experiences in the southern regions.

Digging into the huge and different scenes of southern India, one experiences an embroidery of encounters that unfurl like parts in a dazzling book. From the lavish Western Ghats to the peaceful Deccan Level, the southern locales of India offer a plenty of normal marvels, social wealth, and verifiable profundity that characterize the interesting personality of this dynamic piece of the country.

The Western Ghats, an UNESCO World Legacy Site, stand as a demonstration of the stunning variety of scenes. Extending across a few states in southern India, including Karnataka, Kerala, Tamil Nadu, and Goa, this mountain range is decorated with verdant slopes, thick woods, and flowing cascades. Run Falls in Karnataka, one of the gems of the Western Ghats, presents a scene where the Sharavathi

Stream plunges decisively over a precipice, making a hypnotizing visual orchestra. The geographical wonder of Run Falls is supplemented by its social importance, as it is accepted to be a consecrated site where divinities plunge to favor the land, encouraging an otherworldly association with the normal scene.

Moving toward the south, the province of Kerala reveals the charming Athirapally Falls, frequently alluded to as the "Niagara of India." Settled in the midst of the rich vegetation of the Sholayar Backwoods, Athirapally spellbinds with its flowing waters, making a realistic scene that has been highlighted in different Indian and global movies. The social meaning of Athirapally reaches out to its job as a dream for producers, adding to its prominence among film lovers. The falls, encompassed by native clans and various greenery, typify the amicable conjunction of nature and culture inside the Western Ghats.

Dropping further south, the province of Goa presents the great Dudhsagar Falls, a topographical wonder set against the scenery of the Western Ghats. Named "Ocean of Milk" for the smooth foam made as the water plunges, Dudhsagar unfurls as a scene where the Mandovi Stream dives over a precarious precipice. The falls are not just a geographical miracle but rather hold fanciful importance, related with a princess' requests for virtue. Dudhsagar, wrapped by the Bhagwan Mahavir Untamed life Safe-haven, turns into a journey site where the profound and regular domains meet, displaying the different features of southern India's scenes.

Wandering into the core of the Deccan Level, the scene changes into undulating slopes and extensive levels. Sivasamudram Falls in Karnataka, arranged on the Kaveri Stream, embodies the cooperative energy among water and rock. The falls, fragmented into Barachukki and Gaganachukki, grandstand the geographical development of the Deccan Level, where glasslike rocks make a material for the flowing plunge of the waterway. Past its geographical radiance, Sivasamudram holds authentic significance as one of the earliest hydroelectric power stations in Asia, representing the combination of mechanical development with nature's power.

Venturing to Tamil Nadu, the Shevaroy Slopes present the detached safe house of Kiliyur Falls, settled inside the Yercaud Natural life Asylum. Described by its thin plunge over a rough incline, Kiliyur turns into a safe-haven where biodiversity prospers in the midst of immaculate scenes. The falls, encompassed by thick timberlands, add to the protection of endemic vegetation, featuring the natural meaning of the Deccan Level. Kiliyur, with its fragile excellence, turns into a signal for dependable the travel industry and ecological stewardship inside the southern locales.

Proceeding with the investigation, Hogenakkal Falls in Tamil Nadu unfurls as a progression of fountains and rapids set against the rough bluffs of the Deccan Level. Frequently named the "Niagara of the East," Hogenakkal hypnotizes with its land variety and restorative appeal. The falls are accepted to have mending properties, entwining nature's health with social roots profoundly implanted in conventional practices. Hogenakkal, with its special geographical highlights and social importance, epitomizes the complex encounters that the southern locales of India deal to the individuals who try to investigate its scenes.

As we venture further south, the scenes broaden significantly more, introducing a differentiation between the Deccan Level and the waterfront districts. The southern seaside provinces of Karnataka, Kerala, Tamil Nadu, and Andhra Pradesh brag a rich embroidery of encounters, from quiet sea shores to dynamic urban communities.

The waterfront province of Karnataka is decorated with immaculate sea shores, among which Gokarna stands apart as a serene shelter. Known for its grand excellence and strict importance, Gokarna offers an extraordinary mix of beach front scenes and profound retreats. The sea shores of Gokarna, like Om Ocean side and Kudle Ocean side, give a quiet scenery to dusks and relaxed walks, making an untainted waterfront experience inside the southern domains.

Kerala, frequently alluded to as "God's Own Nation," embraces the Bedouin Ocean with an organization of backwaters, tidal ponds, and coconut-bordered sea shores. The famous backwaters of Alleppey offer a peaceful break, where houseboats journey along the perplexing organization of waterways, displaying the one of a kind way of life of the locale. The social extravagance of Kerala is reflected in its customary dance structures, energetic celebrations, and notable tourist spots, making a powerful waterfront experience that goes past the sandy shores.

The beach front territory of Tamil Nadu presents the clamoring city of Chennai, where advancement meets custom along the Narrows of Bengal. Marina Ocean side, perhaps of the longest metropolitan ocean side on the planet, turns into a center point of movement, facilitating both recreation searchers and local people participated in social merriments. Chennai's energetic expressions scene, verifiable sanctuaries, and pioneer design add to the city's multi-layered character, introducing a beach front encounter that incorporates both metropolitan dynamism and social legacy.

Andhra Pradesh, with its extensive shore along the Straight of Bengal, uncovers the memorable city of Visakhapatnam. The pleasant sea shores of Visakhapatnam, including Ramakrishna Ocean side and Rishikonda Ocean side, furnish a quiet getaway with brilliant sands and sky blue waters. The city's sea history, reflected in its maritime base and submarine exhibition hall, adds a layer of verifiable profundity to the waterfront experience in this southern state.

As we adventure into the southernmost tip of India, the province of Tamil Nadu presents the dazzling town of Kanyakumari. Prestigious for being the juncture of the Bedouin Ocean, the Straight of Bengal, and the Indian Sea, Kanyakumari offers a special waterfront experience where three oceans meet. The Vivekananda Rock Commemoration and Thiruvalluvar Sculpture, set against the scenery of the interminable skyline, become images of profound thought and social solidarity inside the seaside scenes.

The southern districts of India, with their different scenes and encounters, present a kaleidoscope of normal marvels, social lavishness, and verifiable profundity. From the glorious Western Ghats to the peaceful Deccan Level, and from the perfect sea shores of Karnataka to the energetic urban communities along the coast, every objective inside the southern domains offers a special section in the story of this socially and geologically rich piece of the country. As voyagers investigate these scenes, they are not just seeing the magnificence of nature; they are drenching themselves in an embroidery of encounters that characterize the southern locales as a dazzling and diverse objective.

The southern locales of India expand their charm past the scenes, presenting a rich embroidery of social variety and culinary joys that further improve the general insight for voyagers. As one crosses the dynamic southern states, the social subtleties and flavors that characterize every locale become fundamental parts of the excursion.

Social variety in southern India is a living demonstration of the conjunction of different networks, dialects, and customs. Karnataka, with its old sanctuaries and castles, mirrors the verifiable magnificence of lines that once controlled the district. The state's different networks, including Kannadigas, Tulus, and Konkanis, add to a mosaic of dialects, celebrations, and imaginative articulations. The compositional wonders of Hampi and the complex craftsmanship of Mysuru's castles stand as social milestones, welcoming voyagers to dig into the verifiable and creative tradition of Karnataka.

Kerala, suitably named "God's Own Nation," embraces an extraordinary mix of social customs and regular magnificence. The state's performing expressions, like Kathakali and Mohiniyattam, grandstand the energy of Kerala's social legacy. The yearly celebration of Onam, celebrated with enthusiasm, unites networks in a presentation of customary ceremonies, banquets, and people exhibitions. Kerala's culinary variety, set apart by the liberal utilization of coconut, flavors, and fish, adds a delightful aspect to the social experience, making each feast a festival of provincial validness.

Tamil Nadu, with its old sanctuaries and old style expressions, unfurls as a mother lode of social extravagance. The sanctuaries of Madurai, Tanjore, and Kanchipuram stand as engineering wonders, repeating the strict and creative intensity that has characterized the area for quite a long time. Traditional dance structures like Bharatanatyam and Carnatic music add a melodic setting to the social scene of Tamil Nadu. The state's rich scholarly practices, with works like the Sangam verse and Thirukkural, represent the scholarly profundity that saturates Tamil culture.

Andhra Pradesh, with its authentic destinations and lively celebrations, adds to the social mosaic of southern India. The structural splendor of the Venkateswara Sanctuary in Tirupati and the antiquated Buddhist site of Amaravati mirrors the profound and authentic meaning of the locale.

Celebrations like Ugadi and Bhogi are praised with zing, denoting the start of the Telugu New Year. Andhra Pradesh's culinary contributions, known for their red hot flavors and particular flavors, add a gastronomic aspect to the social excursion through the southern states.

The waterfront districts of southern India, with their nearness to the ocean, offer a luscious cluster of fish that entices the taste buds. From the delicious prawns of Kerala's Malabar coast to the hot fish curries of Tamil Nadu and Andhra Pradesh, the beach front cooking is a gastronomic experience in itself. The conventional planning of dishes, frequently went down through ages, adds a social legitimacy to each dinner, welcoming explorers to relish the unmistakable kinds of every district.

Notwithstanding culinary enjoyments, the southern districts are known for their dynamic road food culture. From the fresh dosas and steaming idlis of Karnataka to the tasty biryanis and kebabs of Hyderabad in Telangana, the roads wake up with the smells of assorted and flavorful treats. Each state has its own novel road food contributions, mirroring the neighborhood fixings and culinary practices that have advanced over the long run.

As explorers navigate the southern districts of India, they witness the stunning scenes as well as drench themselves in a social odyssey. The variety of dialects, customs, and foods turns into an embroidery of encounters that improve the excursion, causing it an all encompassing investigation of both the normal and social ponders that to characterize this enthralling piece of the country. From old sanctuaries to vivacious celebrations and delicious dishes, southern India entices with great enthusiasm, welcoming voyagers to leave on a social and culinary caper like no other.

Chapter 6

Conservation and Challenges

Digging into the domain of protection in a worldwide setting, one experiences a perplexing transaction between the safeguarding of normal environments, the practical utilization of assets, and the relieving of anthropogenic effects. This multi-faceted dance between human exercises and the fragile equilibrium of the planet's biodiversity presents both extraordinary open doors and impressive difficulties.

Preservation, at its center, is an undertaking established in the acknowledgment of the characteristic worth of biodiversity and the affirmation that the soundness of environments is intrinsically connected to human prosperity. It encapsulates a promise to defending the rich embroidery of life on The planet, perceiving that every species, regardless of how subtle, adds to the complex trap of environmental communications that supports life.

The essential objective of preservation is to keep up with the biological trustworthiness of normal living spaces and forestall the deficiency of biodiversity. This includes the assurance of jeopardized species, the safeguarding of basic natural surroundings, and the reclamation of environments that have been debased by human exercises. Protection endeavors reach out past the bounds of public lines, perceiving that numerous species set out on transboundary excursions and that worldwide collaboration is fundamental for viable preservation.

One of the foundations of preservation is the foundation and the board of safeguarded regions. These regions, going from public parks to marine stores, act as safe-havens where biodiversity can flourish undisturbed. They go about as sanctuaries for imperiled species, permitting them to recuperate and, at times, forestalling their termination. Safeguarded regions likewise give fundamental biological system administrations, like clean water, carbon sequestration, and environment guideline, helping both the climate and human social orders.

Endeavors to monitor biodiversity reach out past the bounds of safeguarded regions. Preservation science, as a logical discipline, assumes a vital part in understanding the elements of biological systems, the dangers looked by species, and the techniques required for their endurance. Analysts and moderates team up to accumulate information on species dissemination, populace elements, and the effect of human exercises on biological systems. This information frames the reason for informed preservation procedures that address the main drivers of biodiversity misfortune.

The idea of protection isn't restricted to the conservation of appealling species or notable scenes. It incorporates the conservation of hereditary variety inside populaces, perceiving the significance of keeping up with the full range of hereditary variety inside species. This hereditary variety is the natural substance for development, permitting species to adjust to changing ecological circumstances. Preservation endeavors frequently include the ID of key biodiversity regions, focal points of endemism, and passageways that work with the development of species across scenes.

While preservation expects to safeguard and reestablish environments, it additionally embraces the reasonable utilization of regular assets. This methodology, known as practical preservation, perceives that human social orders are unpredictably connected to environments and that mindful asset the executives is fundamental for the prosperity of both nature and individuals. Reasonable preservation includes practices like supportable ranger service, fisheries the executives, and agro-ecology, where farming frameworks impersonate normal biological systems.

Challenges on the Preservation Skyline:

In spite of the respectable goals and huge steps made in the field of protection, various difficulties pose a potential threat not too far off, compromising the fragile equilibrium that supports biodiversity and environments.

Natural surroundings Misfortune and Fracture:

Natural surroundings misfortune and fracture, basically determined by rural extension, urbanization, and foundation advancement, stay among the most basic dangers to biodiversity. As normal scenes are changed over into farming fields, metropolitan regions, or modern zones, species lose their homes and the availability of biological systems is upset.

This imperils the endurance of numerous species as well as subverts the strength of biological systems even with ecological changes.

Environmental Change:

Environmental change, prompted by human exercises like the consuming of non-renewable energy sources and deforestation, represents an existential danger to biodiversity. The adjustment of temperature and precipitation designs, rising ocean levels, and the rising recurrence of outrageous climate occasions straightforwardly influence biological systems and the species they harbor. Numerous species face the test of adjusting to these quick changes, and those unfit to do so may confront eradication.

Overexploitation of Assets:

The overexploitation of normal assets, driven by impractical hunting, fishing, and logging, represents an extreme danger to various species. The interest for untamed life items, like ivory, intriguing pets, and conventional meds, drives poaching and unlawful exchange. Overfishing exhausts marine assets, while logging adds to deforestation. Resolving these issues requires tough guidelines and policing well as supportable options that give vocations without compromising biodiversity.

Intrusive Species:

The presentation of intrusive species, whether deliberately or incidentally, can devastatingly affect local environments. Intrusive species frequently outcompete or go after local species, prompting populace declines and biological system lopsided characteristics. Controlling or destroying obtrusive species is a perplexing test, requiring global joint effort and imaginative administration methodologies.

Contamination:

Contamination, in different structures, represents an unavoidable danger to environments. From substance poisons in water to air contamination influencing respiratory frameworks in untamed life, the effects are extensive. Plastic contamination, specifically, has arisen as a worldwide worry, with microplastics invading marine biological systems and earthbound conditions. The negative impacts of contamination reach out past untamed life to human populaces, stressing the interconnectedness of natural wellbeing.

Sickness Episodes:

Sickness episodes, exacerbated by elements, for example, living space annihilation and environmental change, can disastrously affect natural life populaces. Arising irresistible sicknesses, frequently communicated among natural life and people, feature the requirement for a comprehensive One Wellbeing approach that perceives the interconnected strength of environments, creatures, and individuals.

Absence of Mindfulness and Financing:

Notwithstanding the basic significance of preservation, there is many times an absence of mindfulness and understanding among the overall population. Protection endeavors need monetary help, and getting subsidizing for long haul ventures can be a huge test. Also, contending interests for assets might redirect consideration and assets from protection drives.

Globalization and Exchange:

Globalization has expanded the development of products, including extraordinary species and their illnesses, adding to the spread of intrusive species and arising dangers. Global exchange, while cultivating monetary development, can likewise drive unreasonable asset extraction and untamed life exchange. Tending to these difficulties requires worldwide participation, administrative systems, and manageable exchange rehearses.

Political and Financial Elements:

Political precariousness, absence of administration, and financial factors, for example, neediness can fuel ecological difficulties. In locales where human populaces battle with essential requirements, preservation may not be a main concern. Making an equilibrium that tends to both the necessities of nearby networks and the preservation of biodiversity is fundamental for long haul achievement.

Innovative and Moral Situations:

The utilization of trend setting innovations, like hereditary designing and manufactured science, presents moral difficulties in preservation. While these advances might offer creative arrangements, inquiries regarding unseen side-effects, moral contemplations, and potential biological disturbances should be painstakingly explored.

Tending to these difficulties requires a complex methodology that coordinates logical exploration, strategy mediations, local area commitment, and worldwide cooperation. Preservation endeavors should be versatile, perceiving the powerful idea of environments and the steadily advancing difficulties presented by human exercises and worldwide changes.

The protection of biodiversity remains as a fundamental obligation regarding humankind. It isn't only a question of saving charming species or safeguarding pleasant scenes; it is tied in with guaranteeing the wellbeing and versatility of the planet we call home. The difficulties are overwhelming, however the desperation and significance of the main job request coordinated and supported endeavors from people, networks, states, and the worldwide local area. Just through an aggregate obligation to stewardship and manageability might we at any point desire to get a future where biodiversity flourishes, biological systems prosper, and the sensitive equilibrium of life on Earth perseveres for a long time into the future.

Proceeding with the investigation of preservation challenges, underscoring the interconnectedness of these issues and the requirement for comprehensive, foundational solutions is basic.

Protection endeavors can't be secluded from more extensive cultural and financial settings; they should be incorporated into approaches, practices, and outlooks that focus on maintainability, ecological morals, and the prosperity of the two biological systems and human networks.

Loss of Social Variety:

Preservation moves reach out past natural aspects to incorporate social variety. Native and neighborhood networks frequently have significant biological information and practices that add to biodiversity preservation. Nonetheless, their conventional ways of life are progressively undermined by outside pressures, including land infringement, asset extraction, and uprooting. Perceiving and regarding the freedoms and information on native people groups isn't just fundamental for civil rights yet in addition critical for successful protection. Feasible protection approaches should include and engage nearby networks, coordinating conventional insight with current logical experiences.

Land Use Change and Farming:

Farming, a major human action, assumes a urgent part in molding scenes. Be that as it may, the development of horticultural land frequently comes to the detriment of normal living spaces. Impractical cultivating works on, including monocultures, over the top utilization of agrochemicals, and the change of assorted scenes into enormous scope ranches, add to living space annihilation and biodiversity misfortune. Executing agroecological works on, advancing feasible cultivating strategies, and incorporating preservation measures into agrarian scenes are crucial stages towards offsetting food creation with ecological protection.

Human-Natural life Struggle:

As human populaces extend and infringe upon regular natural surroundings, clashes among people and untamed life heighten. Creatures might strike crops, present dangers to animals, or even imperil human lives. Accordingly, people group might turn to retaliatory measures, further jeopardizing species. Adjusting the necessities of nearby networks with the preservation of untamed life requires creative arrangements, for example, the improvement of untamed life passages, local area based protection drives, and the utilization of innovation to relieve struggle.

Lawful and Administrative Difficulties:

Irregularities and holes in lawful structures present huge difficulties to successful preservation. Natural life dealing, unlawful logging, and unreasonable fishing keep on flourishing because of frail authorization and administrative provisos. Reinforcing global and public regulations, upgrading authorization components, and advancing joint effort between state run administrations, non-legislative associations, and nearby networks are fundamental for handling these difficulties.

Training and Backing:

Building mindfulness and encouraging a feeling of obligation among people is a basic part of effective protection.

Training drives that feature the worth of biodiversity, environmental interdependencies, and the effect of individual decisions can develop a protection mentality. Backing endeavors should connect with assorted partners, from policymakers to nearby networks, to advance arrangements that focus on manageability, preservation, and dependable asset the executives.

Mechanical Headways in Protection:

While innovation offers imaginative instruments for preservation, it likewise presents moral and functional situations. The utilization of robots, satellite symbolism, and hereditary designing can help with observing and research, yet moral contemplations should direct their application. Finding some kind of harmony between innovative progressions and moral obligations is significant to guaranteeing that preservation rehearses stay lined up with the standards of maintainability and regard for environments.

Public and Confidential Area Cooperation:

Coordinated effort between the general population and confidential areas is fundamental for accomplishing practical protection results. Corporate obligation, reasonable strategic approaches, and associations among organizations and protection associations can add to positive ecological effects. Drawing in the confidential area in protection drives, for example, reforestation projects, territory rebuilding, and maintainable asset the board, can adjust financial interests to biological stewardship.

Versatility and Versatile Administration:

Building strength in environments and embracing versatile administration procedures are vital to preservation achievement. Perceiving that environments are dynamic and dependent upon future developments, preservation approaches should be adaptable and responsive. This incorporates recognizing the job of aggravation, like fierce blazes or normal progression, in forming environments. Coordinating strength thinking into protection arranging considers techniques that can adjust to changing natural circumstances.

Worldwide Cooperation and Tact:

Protection challenges frequently rise above public boundaries, requiring worldwide coordinated effort. Peaceful accords, like the Show on Natural Variety (CBD), give systems to collaboration. Nonetheless, more prominent conciliatory endeavors and responsibilities are expected to resolve issues, for example, deforestation, environmental change, and the unlawful untamed life exchange. A common perspective of the interconnectedness of biological systems and biodiversity can encourage worldwide collaboration to safeguard the planet's normal legacy.

Local area Strengthening and Supportable Jobs:

Preservation endeavors should focus on local area strengthening and the making of economical jobs.

Coordinating protection rehearses with nearby economies, like ecotourism or supportable reaping of woodland items, can adjust the interests of networks to the objectives of biodiversity preservation. This approach perceives that the prosperity of individuals and the wellbeing of biological systems are interlaced, encouraging a mutually beneficial situation for both.

Despite these difficulties, the basic for activity is clear. An extensive and coordinated way to deal with protection is fundamental, one that thinks about biological, social, and financial aspects. This requires not just the responsibility of states, preservation associations, and researchers yet additionally the dynamic support of networks, organizations, and people.

The way ahead includes encouraging a worldwide ethic that qualities and regards the variety of life, perceiving that protection isn't an extravagance yet a crucial need for the proceeded with endurance and success of mankind. By tending to these difficulties with a unified front, embracing feasible practices, and developing a profound appreciation for the interconnectedness of every single residing being, we can endeavor towards a future where biodiversity flourishes, biological systems prosper, and the sensitive equilibrium of the planet is protected for a long time into the future.

6.1 Address the environmental challenges faced by India's waterfalls.

India's cascades, glorious fountains that typify the country's normal magnificence, are not insusceptible to the ecological difficulties that influence biological systems around the world. As famous milestones and significant parts of India's different scenes, cascades face a range of dangers that envelop both normal and anthropogenic elements. Understanding and tending to these difficulties is essential for the protection of these normal marvels and the fragile environments that encompass them.

One of the principal natural difficulties looked by India's cascades is territory adjustment and corruption. The regions encompassing cascades frequently act as basic territories for different widely varied vegetation, making one of a kind environments adjusted to the dampness and microclimates produced by the falling water. Human exercises like deforestation, urbanization, and agrarian extension can prompt living space misfortune and discontinuity, upsetting the sensitive equilibrium of these biological systems. The transformation of regular scenes into farming fields or metropolitan regions not just decreases the stylish worth of the cascades yet in addition represents an immediate danger to the biodiversity they support.

Deforestation, driven by wood extraction, farming extension, and framework improvement, is a huge supporter of natural surroundings misfortune around cascades. The getting free from timberlands not just decreases the picturesque magnificence of the cascade environmental elements yet additionally disturbs the biological elements of the woods environment.

Woods assume a pivotal part in directing water stream, forestalling soil disintegration, and giving environment to various species. The deficiency of these capabilities can have flowing consequences for the strength of the cascade biological systems.

Urbanization, frequently prodded by the travel industry and formative ventures, represents another test. The development of inns, resorts, and framework close to cascades can prompt living space discontinuity and change of regular waste examples. Also, the deluge of vacationers might bring about expanded contamination, aggravation to natural life, and the presentation of intrusive species, further affecting the environmental respectability of the cascade biological systems.

Rural development nearby cascades can add to soil disintegration and water defilement. Overflow from agrarian regions might convey silt and agrochemicals into water bodies, influencing water quality downstream. These toxins can inconveniently affect amphibian life, disturbing the fragile equilibrium of the environments related with the cascades.

Environmental change addresses a worldwide test that is intensely influencing cascades in India. Changes in temperature and precipitation examples can modify the hydrological cycles, affecting the progression of waterways and streams that feed the cascades. Changes in precipitation might prompt varieties in the volume of water flowing over the falls, affecting their visual exhibition and natural capabilities. Moreover, climbing temperatures can influence the dispersion and conduct of plant and creature species in cascade biological systems, possibly prompting shifts in biodiversity.

One of the apparent effects of environmental change on cascades is the modification of occasional stream designs. Changes in precipitation designs and the planning of rainstorm can impact the water levels, influencing the power of the cascades during various seasons. This changeability can have biological ramifications for species that depend on the occasional elements of the cascades for rearing, scavenging, or other life cycle occasions.

Past normal variables, anthropogenic exercises fundamentally add to the debasement of water quality around cascades. Contamination from farming spillover, modern releases, and ill-advised garbage removal can pollute the water bodies related with cascades. The amassing of poisons can hurt sea-going organic entities, upset food networks, and compromise the general wellbeing of the environments.

The travel industry, while giving financial advantages to nearby networks, can likewise present natural difficulties. Unregulated the travel industry can prompt territory aggravation, littering, and the stomping on of delicate vegetation. The aggregate effect of enormous quantities of guests can debase the stylish allure of cascades and upset the regular way of behaving of untamed life nearby. It is significant to find some kind of harmony between advancing the travel industry and carrying out feasible practices that limit the environmental impression of guests.

The presentation of obtrusive species is one more ecological test looked by cascades. Human exercises, including the travel industry and urbanization, can incidentally acquaint non-local species with cascade biological systems. These intrusive species may outcompete local verdure, disturb biological cycles, and change the construction of the environments. At times, obtrusive species might try and represent an immediate danger to local species, compounding the weakness of cascade environments.

Hydroelectric activities and dam development's address extra difficulties to cascade environments. While these undertakings intend to tackle water assets for energy creation and different purposes, they can have huge biological results. Adjustments to stream, silt transport, and water temperature can influence the territories around cascades and downstream regions. Dams may likewise block the normal movement of fish and other oceanic creatures, upsetting the natural equilibrium of stream biological systems.

The combined effect of these natural difficulties requires a thorough and incorporated way to deal with the preservation of India's cascades. Drives should address the underlying drivers of debasement, advance maintainable practices, and include nearby networks in the stewardship of these normal marvels.

Preservation Systems:

Environment Rebuilding and Insurance:

Carrying out territory reclamation projects around cascades is vital for switching the effects of deforestation and natural surroundings debasement. This includes re-planting local vegetation, reestablishing regular waste examples, and shielding basic living spaces from additional infringement. Laying out support zones and safe-guarded regions around cascades can add to the protection of these environments.

Manageable The travel industry Practices:

Overseeing the travel industry in a manageable way is fundamental for adjusting the financial advantages of the travel industry with the protection of cascade environments. Executing guest rules, laying out conveying limits, and advancing eco-accommodating practices can limit the biological impression of the travel industry. Local area based the travel industry drives that include nearby networks in the dynamic cycle can guarantee that monetary advantages line up with protection objectives.

Environmental Change Relief and Variation:

Tending to the effects of environmental change on cascades requires a double methodology of moderation and transformation. Moderation endeavors include diminishing ozone harming substance emanations to restrict further environmental change, while transformation techniques center around building the versatility of cascade biological systems to evolving conditions.

This might incorporate reestablishing riparian vegetation, carrying out water protection measures, and checking biological reactions to environment change-ability.

Water Quality Administration:

Safeguarding the water nature of waterways and streams related with cascades is essential for the strength of oceanic biological systems. Carrying out measures to decrease farming overflow, controlling modern releases, and advancing waste administration practices can moderate water contamination. Customary observing of water quality boundaries is fundamental for early location of contamination and brief remedial activity.

Intrusive Species Control:

Creating and carrying out obtrusive species the board plans is vital for safe-guarding the trustworthiness of cascade biological systems. This might include the evacuation of intrusive species, reclamation of local vegetation, and the foundation of isolation measures to forestall the presentation of new obtrusive species. Public mindfulness missions can instruct networks and guests about the likely effects of obtrusive species and how to forestall their spread.

Hydroelectric Undertaking Arranging:

The preparation and execution of hydroelectric activities should think about the natural ramifications on cascade biological systems. Directing intensive natural effect evaluations, taking on fish-accommodating plans, and executing measures to alleviate living space interruption are fundamental stages. Adjusting the require ment for energy creation with the protection of sea-going environments is a critical thought in practical water the executives.

Local area Commitment and Schooling:

Drawing in nearby networks in protection endeavors is essential to the progress of drives around cascades. Teaching people group about the environmental significance of cascades, encouraging a feeling of stewardship, and including them in dynamic cycles add to long haul supportability. Local area based protection projects that give elective occupations and engage nearby inhabitants as preservation envoys can upgrade the adequacy of protection techniques.

Regulation and Authorization:

Reinforcing ecological regulation and requirement systems is basic for the insurance of cascade environments. Hearty legitimate structures, upheld by viable requirement, discourage criminal operations like deforestation, contamination, and unapproved development. Joint effort between government offices, non-legislative associations, and nearby networks is fundamental for guaranteeing consistence with natural guidelines.

Examination and Checking:

Progressing exploration and observing endeavors are essential for figuring out the elements of cascade environments and evaluating the viability of preservation systems.

This incorporates concentrating on the natural connections, observing water quality boundaries, and following changes in biodiversity. The information created through research illuminate proof based preservation choices and assist with adjusting methodologies to advancing ecological circumstances.

Global Coordinated effort:

Cascade biological systems frequently length global limits, accentuating the requirement for cross-line coordinated effort. Sharing information, best practices, and assets with adjoining nations can improve the aggregate ability to address transboundary ecological difficulties. Peaceful accords and associations can work with cooperative endeavors in exploration, preservation, and supportable administration of shared cascade biological systems.

Proceeding with the investigation of ecological difficulties looked by India's cascades, it is pivotal to dive further into the complicated interchange of variables that add to the weakness of these regular miracles. The protection systems illustrated before give an establishment to tending to these difficulties, however a nuanced comprehension of explicit issues and their local varieties is fundamental for viable and customized intercessions.

Disintegration and Sedimentation:

Disintegration in the catchment areas of cascades represents a critical danger to their strength and environmental wellbeing. Deforestation, horticultural practices, and development exercises can speed up soil disintegration, prompting an expanded silt load in waterways and streams that feed the cascades. Extreme sedimentation can modify the stream elements, corrupt water quality, and cover amphibian living spaces. Protection endeavors should incorporate measures to alleviate disintegration, for example, reforestation, soil preservation rehearses, and the foundation of support zones to limit dregs spillover.

Biodiversity Misfortune:

The biodiversity related with cascade environments is unpredictably connected to the accessibility of different living spaces, stable hydrological conditions, and a solid encompassing scene. Territory debasement and discontinuity, contamination, and environmental change add to the deficiency of biodiversity. Endemic species, which are frequently remarkably adjusted to cascade conditions, face elevated weakness. Preservation techniques ought to focus on the security of basic environments, the rebuilding of corrupted regions, and the execution of measures to defend endemic and compromised species.

Social and Profound Importance:

Numerous cascades in India hold social and profound importance for nearby networks, frequently being respected as consecrated destinations. Expanded human exercises, the travel industry, and urbanization can disturb the customary and otherworldly practices related with these cascades.

Preservation drives shouldn't just expect to safeguard the environmental trustworthiness of cascade biological systems yet additionally regard and coordinate the social qualities and convictions of nearby networks. Cooperative endeavors that include local area pioneers, strict establishments, and protection associations can assist with finding some kind of harmony between social practices and biological conservation.

Microplastic Contamination:

An arising ecological concern is the presence of microplastics in water bodies around cascades. Plastic contamination, coming about because of inappropriate garbage removal and the breakdown of bigger plastic things, has invaded even far off common habitats. Microplastics can negatively affect amphibian organic entities, upset food networks, and posture expected dangers to human wellbeing. Preservation systems ought to incorporate measures to diminish plastic use, carry out legitimate waste administration practices, and direct standard observing for microplastic defilement in cascade environments.

Water Shortage and Changed Hydrology:

Environmental change and anthropogenic exercises can modify the hydrological examples of streams and streams, affecting the water accessibility for cascades. Changes in precipitation, deforestation, and the redirection of water for farming and modern purposes can add to water shortage. Preservation endeavors ought to resolve the more extensive issues of water asset the board, advancing maintainable practices that focus on natural necessities while fulfilling the needs of human social orders. Keeping up with normal hydrological cycles is indispensable for saving the tasteful and natural elements of cascades.

Financial Tensions:

The regions encompassing cascades frequently experience financial tensions, driven by elements like populace development, destitution, and absence of elective vocation choices. These tensions can compound ecological corruption as networks exploit normal assets for endurance. Preservation systems need to consolidate financial advancement drives that give elective occupations, advance economical asset the executives, and engage neighborhood networks as stewards of their regular legacy.

Legitimate Systems and Administration:

Deficient legitimate systems and administration designs can block viable protection. Reinforcing natural regulations, guaranteeing their implementation, and laying out clear rules for land use arranging are basic advances. Cooperative administration models that include neighborhood networks, legislative organizations, and non-administrative associations can upgrade the viability of protection drives. Straightforwardness and responsibility in dynamic cycles are fundamental for building trust and guaranteeing the drawn out progress of protection endeavors.

Instruction and Mindfulness:

Public mindfulness and schooling assume a critical part in cultivating a feeling of obligation and protection morals. Instructive projects ought to target neighborhood networks, schools, and sightseers, underlining the significance of cascades, their natural importance, and the job people can play in their safeguarding. Interpretative signage, directed visits, and effort projects can upgrade public comprehension and enthusiasm for cascade biological systems.

Research and Versatile Administration:

Progressing logical exploration is basic for figuring out the powerful idea of cascade environments and adjusting protection procedures to evolving conditions. Research drives ought to zero in on the hydrology, biodiversity, and natural cycles related with cascades. Versatile administration draws near, informed by logical experiences, take into consideration the persistent refinement of preservation systems in light of developing ecological difficulties and reactions.

Transboundary Preservation:

Numerous cascades in India length global limits, requiring cooperative protection endeavors with adjoining nations. Laying out transboundary preservation arrangements, sharing information and ability, and planning the board rehearses are fundamental for the comprehensive security of cascade environments. Conciliatory drives and local organizations can add to the supportable administration of shared normal assets.

The ecological difficulties looked by India's cascades are diverse and interconnected, requesting complete and setting explicit preservation draws near. As these regular marvels keep on attracting admirers and assume indispensable parts supporting biological systems, the earnestness of proactive and versatile preservation methodologies couldn't possibly be more significant. By tending to the complicated trap of natural, social, and social variables, India can shield cascades as versatile and flourishing biological systems persevere for a long time into the future. The obligation to protection isn't simply an obligation to the climate; it is a promise to safeguard the natural magnificence, environmental respectability, and social meaning of these flowing miracles that effortlessness the different scenes of the country.

6.2 Explore conservation efforts and initiatives to protect these natural wonders.

The protection of normal marvels, including cascades, is a vital undertaking that requires purposeful endeavors at neighborhood, public, and worldwide levels. Perceiving the natural, social, and stylish meaning of cascades, different preservation drives have been started to safeguard these remarkable environments and the encompassing scenes. These endeavors envelop a scope of procedures and activities pointed toward protecting biodiversity, moderating anthropogenic effects, and encouraging feasible connections between human networks and common habitats.

Safeguarded Regions and Save Assignments:

One of the basic methodologies in cascade preservation is the foundation of safeguarded regions and save assignments. Public parks, untamed life safe-havens, and biosphere saves assume an essential part in protecting cascade biological systems. These regions are assigned to save the remarkable biodiversity, keep up with normal cycles, and give undisturbed living spaces to verdure related with cascades. Severe guidelines inside these safeguarded regions limit human exercises, guaranteeing insignificant unsettling influence to the sensitive biological systems.

Reforestation and Territory Rebuilding:

Addressing deforestation and territory corruption is basic to cascade preservation. Reforestation endeavors include establishing local tree species in regions that have been cleared or corrupted. These drives not just add to the recuperation of regular environments yet in addition help in settling soil, forestalling disintegration, and further developing water quality. Territory rebuilding projects, frequently attempted in a joint effort with neighborhood networks, expect to reproduce the circumstances helpful for the flourishing biodiversity related with cascades.

Feasible The travel industry Practices:

The travel industry, while adding to neighborhood economies, can likewise present huge dangers to cascade biological systems. Unregulated the travel industry can bring about territory aggravation, littering, and contamination. Reasonable the travel industry rehearses intend to adjust the financial advantages of the travel industry with the safeguarding of indigenous habitats. This incorporates the execution of guest rules, conveying limit appraisals, and eco-accommodating framework to limit the environmental impression of travelers. Local area based the travel industry drives engage nearby networks and give financial motivators to the protection of cascades.

Local area Commitment and Strengthening:

Including neighborhood networks in protection endeavors is major for long haul achievement. Perceiving the reliance of networks on normal assets, protection drives endeavor to engage neighborhood inhabitants as stewards of their current circumstance. Local area commitment includes cooperative navigation, instruction programs, and the improvement of elective livelihoods that are lined up with feasible asset the executives. By coordinating the information and desires of neighborhood networks, protection endeavors become more comprehensive and viable.

Instruction and Mindfulness Projects:

Bringing issues to light about the biological significance of cascades is urgent for gathering public help and cultivating a protection outlook. Instructive projects target schools, nearby networks, and vacationers, giving data about the meaning of cascades, the biological systems they support, and the possible effects of human exercises.

Interpretative signage, directed visits, and effort programs add to improving public comprehension and enthusiasm for cascade conditions.

Logical Exploration and Observing:

Logical examination fills in as the establishment for proof based preservation systems. Specialists concentrate on the hydrology, biodiversity, and environmental cycles related with cascades to acquire experiences into their elements. Checking programs track changes in water quality, species populaces, and natural surroundings conditions over the long run. This nonstop progression of logical data illuminates versatile administration works on, permitting protection drives to advance in light of the most recent exploration discoveries.

Obtrusive Species The executives:

The presentation of obtrusive species represents a huge danger to cascade biological systems. Obtrusive plants, creatures, or microbes can outcompete local species, upset natural cycles, and adjust the design of the environment. Preservation endeavors incorporate the turn of events and execution of obtrusive species the board plans. These plans might include the evacuation of intrusive species, natural surroundings rebuilding, and preventive measures to check the presentation of new obtrusive species.

Regulation and Strategy Improvement:

Strong legitimate systems and strategies are fundamental for the compelling protection of cascades. Legislatures sanction regulations that direct exercises in and around cascade environments, resolving issues like deforestation, contamination, and natural surroundings obliteration. Strategy improvement includes a cooperative interaction that connects with different partners, including government offices, non-legislative associations, and neighborhood networks. Clear rules and implementation systems guarantee consistence with protection guidelines.

Worldwide Joint effort:

Perceiving that numerous cascades length global limits, it is basic to cultivate coordinated effort between nations. Transboundary preservation drives include sharing data, organizing the board rehearses, and tending to normal difficulties. Peaceful accords and organizations add to a comprehensive way to deal with protection that thinks about the interconnected idea of cascade biological systems. Cooperative endeavors likewise reinforce conciliatory ties and cultivate a feeling of shared liability regarding the insurance of these normal miracles.

Environmental Change Transformation Techniques:

As environmental change presents difficulties to the hydrological designs and natural elements of cascades, preservation endeavors should consolidate transformation procedures. These may incorporate measures to reestablish riparian vegetation, improve watershed strength, and screen the effects of environmental change on species dispersions.

By coordinating environmental change contemplations into protection arranging, drives can be more receptive to the advancing circumstances influencing cascade biological systems.

Corporate and Confidential Area Commitment:

The corporate and confidential areas can assume a fundamental part in cascade protection. Organizations with activities close to cascade biological systems might add to preservation through dependable strategic policies, practical asset the board, and backing for nearby networks. Corporate associations with preservation associations, sponsorship of exploration drives, and adherence to natural norms add to the more extensive objective of maintainable turn of events.

Charitable Help and Subsidizing:

Preservation drives need monetary help for research, living space rebuilding, local area commitment, and implementation endeavors. Humanitarian associations, establishments, and people contribute financing to help cascade protection projects. Awards, gifts, and sponsorship amazing open doors empower protection associations to carry out complete systems and address the assorted difficulties looked by cascade environments.

Versatile Administration and Adaptability:

Perceiving the unique idea of biological systems and the vulnerabilities related with ecological changes, protection drives embrace versatile administration standards. Versatile administration includes an adaptable and iterative methodology that takes into consideration consistent learning and change of techniques in view of observing information and criticism. This versatility based approach guarantees that protection endeavors stay successful notwithstanding advancing difficulties.

Protection Through Economical Turn of events:

Coordinating protection rehearses with economical improvement objectives is fundamental for accomplishing an agreeable harmony between human necessities and natural conservation. Supportable improvement drives advance practices that address the issues of present ages without compromising the capacity of people in the future to address their own issues. Protection endeavors that line up with economical improvement add to the drawn out prosperity of the two environments and networks.

The preservation of cascades requires a diverse and cooperative methodology that addresses biological, social, monetary, and social aspects. By joining safeguarded region assignments, local area commitment, reasonable the travel industry practices, and global joint effort, protection drives can make an all encompassing structure for saving these stunning normal miracles. As overseers of the planet, it is our common obligation to defend the characteristic worth of cascades, guaranteeing that they keep on motivating amazement, support different environments, and act as images of the fragile harmony among mankind and the normal world.

6.3 Discuss the role of local communities and authorities in preserving the beauty of the cascades.

The protection of fountains, with their stunning excellence and biological importance, depends intensely on the dynamic interest of nearby networks and specialists. Perceiving that these regular miracles are much of the time implanted inside the texture of neighborhood scenes, the job of the people who possess and administer these regions is essential in molding the preservation and practical administration techniques for overflows. From native information to cooperative administration, the contribution of nearby networks and specialists is essential in guaranteeing the drawn out security of these dazzling elements.

Native Information and Customary Practices:

Neighborhood people group, particularly those with native connections to the land, hold an abundance of information about overflows that goes past what logical examination might catch. Native information envelops a comprehension of the occasional examples, biodiversity, and social meaning of these fountains. Customary practices for reasonable asset use, remembering reaping specific plants or looking for a way that regards environmental equilibrium, add to the safeguarding of outpouring biological systems. Perceiving and coordinating this native insight into preservation endeavors is fundamental for all encompassing and socially delicate ways to deal with protecting the magnificence of the fountains.

Social Importance and Stewardship:

Flows frequently hold social importance for nearby networks, becoming essential components of fables, ceremonies, and customs. The profound associations that networks have with these regular highlights ingrain a feeling of stewardship. Nearby inhabitants frequently view themselves as caretakers of the fountains, answerable for their protection and prosperity. Utilizing this social security in preservation endeavors cultivates a feeling of responsibility and pride, empowering networks to partake in safeguarding the stylish and environmental respectability of the fountains effectively.

Local area Based Protection Drives:

Local area based protection drives enable nearby occupants to play a functioning job in the administration of fountains. These drives include cooperative dynamic cycles that consolidate the viewpoints and needs of local area individuals. Laying out neighborhood preservation advisory groups, participating in participatory planning activities, and directing normal local gatherings are instances of comprehensive practices. Through these drives, networks become fundamental accomplices in planning and carrying out protection methodologies that line up with their necessities and desires.

Reasonable Occupations and Monetary Other options:

Offsetting protection objectives with the financial necessities of nearby networks is a key test. Numerous people group living close to overflows rely upon the encompassing regular assets for their vocations. Practical work programs, for example, eco-accommodating the travel industry, agroforestry, or handiworks, can give elective types of revenue that don't think twice about wellbeing of outpouring biological systems. By making monetary impetuses for protection, these drives guarantee that nearby networks view the conservation of the fountains as vital to their financial prosperity.

Instruction and Limit Building:

Enabling nearby networks requires giving them the information and abilities important for successful protection. Instructive projects on overflow nature, maintainable asset the executives, and the significance of biodiversity add to building the limit of local area individuals. Studios, instructional meetings, and mindfulness crusades cultivate a more profound comprehension of the fountains and their part in keeping up with the general strength of environments. Furnishing people group with this information empowers them to settle on informed choices and effectively partake in protection endeavors.

Cooperative Administration Models:

Successful preservation frequently includes joint effort between nearby networks and different degrees of government. Cooperative administration models unite neighborhood specialists, legislative offices, non-legislative associations, and local area agents to oversee overflow biological systems aggregately. Laying out co-administration arrangements, where obligations and dynamic powers are shared, guarantees that the points of view of neighborhood networks are incorporated into protection approaches. This cooperative methodology upgrades the adequacy and supportability of protection drives.

Neighborhood Specialists and Strategy Execution:

Neighborhood specialists assume a basic part in the requirement of protection strategies and guidelines. They are liable for administering land use arranging, observing criminal operations, and executing measures to safeguard overflow environments. Vigorous arrangements, upheld by neighborhood specialists, hinder hurtful practices like deforestation, contamination, or unlawful fishing. The viability of protection arrangements is dependent upon the responsibility and limit of nearby specialists to authorize them.

Compromise and Intervention:

In regions with contending interests, clashes might emerge over asset use and protection needs. Neighborhood specialists act as middle people, assisting with settling clashes and figure out some shared interest between different partners, including networks, organizations, and natural associations. Laying out instruments for compromise guarantees that protection endeavors continue without a hitch, cultivating coordinated effort instead of dissension.

Framework Improvement and Natural Effect Evaluation:

Neighborhood specialists frequently assume a huge part in supporting and managing foundation improvement close to overflows. Directing intensive ecological effect appraisals (EIAs) prior to endorsing undertakings like streets, resorts, or hydropower plants is fundamental. Nearby specialists should offset advancement needs with the protection of outpouring environments. Carrying out rules that focus on ecological maintainability being developed ventures protects the fountains from impeding effects.

The travel industry The board Techniques:

The travel industry, while giving monetary advantages, can present difficulties to flow biological systems in the event that not oversaw reasonably. Nearby specialists are instrumental in creating and authorizing the travel industry the board methodologies. Executing guest rules, setting conveying limits, and managing foundation improvement around overflows are inside the domain of nearby specialists. Feasible the travel industry rehearses guarantee that the deluge of guests upgrades as opposed to compromises the excellence and biological soundness of the fountains.

Water Asset The executives:

In regions where fountains are taken care of by waterways or streams, water asset the executives is a basic part of preservation. Nearby specialists are answerable for dispensing water freedoms, forestalling unlawful water extraction, and guaranteeing that water streams are kept up with to help overflow biological systems. Adjusting the necessities of agribusiness, industry, and networks with the biological prerequisites of the fountains requires educated and impartial water asset the executives.

Catastrophe Readiness and Reaction:

Overflows are defenseless to cataclysmic events like floods, avalanches, or dry seasons. Neighborhood specialists assume a crucial part in creating calamity readiness designs and answering really to crises. Early admonition frameworks, clearing techniques, and post-catastrophe recuperation endeavors are fundamental parts of guaranteeing the versatility of outpouring biological systems and the wellbeing of adjacent networks.

Research Coordinated effort with Neighborhood Organizations:

Coordinated effort with neighborhood instructive organizations and examination focuses upgrades the logical comprehension of outpouring environments. Neighborhood specialists can work with research organizations, supporting investigations on overflow hydrology, biodiversity, and environmental elements. Incorporating neighborhood information into research drives reinforces the logical reason for protection techniques and encourages a feeling of responsibility among networks.

Advancement of Natural Morals:

Nearby specialists can add to the advancement of ecological morals inside networks. Consolidating natural training in school educational plans, coordinating mindfulness crusades, and celebrating nearby biological celebrations impart a feeling of obligation and regard for overflow environments. Ecological morals act as an establishment for maintainable practices and local area drove protection drives.

Transformation to Environmental Change:

Given the effects of environmental change on overflow biological systems, neighborhood specialists are instrumental in creating and carrying out variation procedures. These may incorporate measures to address changing precipitation designs, temperature variances, and modified hydrological cycles. Restricted procedures that consider the particular weaknesses of outpouring environments add to their strength despite environmental change.

Legitimate Systems and Support:

Nearby specialists assume a part in upholding major areas of strength for structures that safeguard overflow biological systems. They can team up with more elevated levels of government to advocate for the institution and authorization of preservation regulations. Taking part in lawful backing guarantees that overflow biological systems get the legitimate security important for their protection.

The conservation of the excellence of fountains is a common obligation that includes the dynamic commitment of both nearby networks and specialists. The perplexing snare of biological, social, and monetary interdependencies requires cooperative endeavors to guarantee that fountains keep on moving stunningness and add to the general soundness of environments. By encouraging comprehensive protection draws near, regarding native information, and embracing maintainable practices, neighborhood networks and specialists become the watchmen of these normal marvels, guaranteeing that their excellence perseveres for a long time into the future.

Chapter 7

The Future Flow

The future progression of fountains, those entrancing regular ponders that out-pouring down rock faces, lavish scenes, and pleasant territories, is complicatedly attached to the developing elements of biological, social, and climatic variables. As we peer into the distance, mulling over the direction that these fountains will take, it becomes clear that the future stream is formed by a fragile dance between protection endeavors, human collaborations, and the effects of an evolving environment. This investigation digs into the different aspects that characterize the eventual fate of fountains, crossing biological versatility, reasonable practices, social safeguarding, and the basic to adjust to a world going through significant natural changes.

Biological Strength Notwithstanding Environmental Change:

Environmental change represents a considerable test to the future progression of fountains. Modifications in precipitation designs, temperature varieties, and outrageous climate occasions can essentially affect the hydrological cycles that support these regular marvels. Overflows, with their reliance on steady water stream, face the gamble of disturbed occasional examples, unpredictable water accessibility, and changes in generally biological system elements. The way to guaranteeing the future flexibility of fountains lies in versatile systems that relieve the effects of environmental change and improve the natural vigor of these biological systems.

Protection drives should coordinate environmental change contemplations into their systems. This includes checking and demonstrating the possible effects of environmental change on overflow environments, recognizing weak species and territories, and executing measures to upgrade the versatile limit of these normal frameworks. Reclamation extends that emphasis on restoring local vegetation, safeguarding watershed regions, and advancing biodiversity assume a urgent part in building biological strength. Additionally, the foundation of environment strong passageways can work with the development of species, permitting them to adjust to changing natural circumstances.

Supportable Practices for Proceeded with Excellence:

Reasonable practices are the key part for keeping up with the inborn excellence of fountains. The fragile biological systems encompassing these regular miracles are many times delicate to human exercises, and the reception of supportable practices is fundamental to forestall irreversible corruption. This includes a range of activities, from mindful the travel industry to eco-accommodating foundation improvement, which looks to work out some kind of harmony between human pleasure and biological conservation.

The travel industry, a situation with two sides, can be a wellspring of income for neighborhood networks while likewise presenting dangers to flow biological systems. Feasible the travel industry rehearses accentuate low-influence appearance, adherence to assigned trails, and the execution of conveying limit cutoff points to forestall over-the travel industry. Interpretative projects that teach guests about the delicacy of outpouring environments and the significance of dependable way of behaving add to cultivating a preservation disapproved of ethos among sightseers.

Foundation improvement nearby fountains should stick to rigid natural principles. Eco-accommodating development works on, including the utilization of privately obtained materials and negligible natural surroundings disturbance, assist with relieving the biological impression of human turns of events. Planning framework that supplements the normal feel of fountains, for example, raised walkways or perception stages, improves guest encounters without compromising the honesty of the biological systems.

Social Conservation and Nearby Stewardship:

Flows frequently hold significant social importance for neighborhood networks, woven into the embroidered artwork of their customs, ceremonies, and conviction frameworks. The safeguarding of this social legacy is a basic part of getting the future progression of fountains. Nearby stewardship, established as it were of social pride and obligation, arises as a strong power in supporting these regular marvels.

Drawing in neighborhood networks as stewards of outpouring biological systems includes recognizing and regarding their social associations with these scenes. Preservation drives ought to effectively include local area individuals in dynamic cycles, guaranteeing that their points of view, conventional information, and yearnings are coordinated into the board systems.

Laying out local area drove preservation programs, upheld by organizations with legislative and non-administrative substances, enables nearby occupants to play a functioning job in protecting the social and biological respectability of fountains.

Social safeguarding endeavors likewise reach out to training and mindfulness programs that commend the rich legacy related with overflows. Schools, public venues, and social organizations become central focuses for dispersing information about the verifiable meaning of these regular ponders, the tales implanted in nearby legends, and the job of fountains in molding local area personality. By cultivating a feeling of social pride and association, preservation drives motivate an aggregate obligation to protecting the legacy of fountains for people in the future.

Versatile Systems for Evolving Scenes:

The scenes encompassing fountains are not static; they are likely to advancing area use designs, segment shifts, and financial changes. Versatile systems are basic to explore the intricacies of these unique scenes and guarantee the supported progression of fountains into what's in store. This includes a proactive way to deal with address arising difficulties and exploit potential open doors for preservation and feasible administration.

Land use arranging arises as an essential apparatus in directing the turn of events and security of fountain environments. Cooperative endeavors between nearby specialists, natural organizations, and local area agents can bring about the plan of land use arrangements that focus on the preservation of basic territories, limit unsafe exercises, and lay out cradle zones around overflows. Versatile administration rehearses, informed by progressing exploration and observing, empower protection methodologies to be receptive to changing natural circumstances and arising dangers.

Mix of fountains into more extensive scene protection drives is fundamental for keeping up with network between biological systems. Laying out biological halls that connection overflow living spaces with encompassing wild regions works with the development of species, upholds hereditary variety, and upgrades in general environment versatility. Traditionalists and policymakers should work cooperatively to distinguish key hallways, address scene discontinuity, and carry out measures that permit fountains to flourish as essential parts of bigger environmental organizations.

Mechanical Developments for Observing and Exploration:

Progressions in innovation offer phenomenal open doors for checking overflow biological systems, directing examination, and executing viable protection techniques. Remote detecting innovations, like satellite symbolism and automated airborne vehicles (UAVs), give important apparatuses to surveying scene changes, observing vegetation wellbeing, and recognizing expected dangers to flow environments.

These innovations upgrade the proficiency and exactness of information assortment, empowering progressives to come to informed conclusions about asset distribution and mediation methodologies.

Besides, the combination of Geographic Data Framework (GIS) planning takes into consideration the spatial investigation of fountain territories, recognizable proof of biodiversity areas of interest, and evaluation of scene network. These devices help with recognizing need regions for protection, outlining support zones, and planning systems to relieve the effect of human exercises on overflow environments.

Mechanical developments likewise assume a critical part in research tries zeroed in on grasping the complicated elements of outpouring biological systems. Progressed hydrological displaying, hereditary investigations, and environment demonstrating add to a more profound comprehension of fountain flexibility, species communications, and the more extensive biological cycles that support these regular miracles. The information produced through mechanical intercessions frames the reason for proof based protection rehearses that think about the mind boggling snare of biological collaborations inside overflow environments.

Worldwide Coordinated effort for Transboundary Protection:

Many fountains length worldwide limits, requiring cooperative endeavors between nations to guarantee their compelling preservation. Global joint effort is instrumental in tending to transboundary challenges, sharing prescribed procedures, and cultivating a common feeling of obligation for the conservation of these interconnected environments.

Laying out transboundary preservation arrangements and associations works with composed endeavors between adjoining nations. These arrangements might incorporate components for data trade, joint exploration drives, and cooperative administration techniques. By rising above international limits, global joint effort adds to an all encompassing methodology that thinks about the whole biological setting of fountains, perceiving the significance of shared liabilities in saving these regular marvels.

Conciliatory drives, upheld by natural associations and legislative organizations, assume a vital part in building agreement on protection needs and arranging arrangements that focus on the environmental wellbeing of fountain biological systems. Global joint effort encourages a feeling of collaboration, common comprehension, and aggregate activity, rising above political contrasts for the common objective of rationing overflows for present and people in the future.

Schooling as an Impetus for Change:

Schooling arises as a strong impetus for driving positive change in the preservation and feasible administration of fountains. Designated instructive projects, spreading over schools, colleges, and local area outreach drives, are instrumental in forming perspectives, encouraging ecological morals, and imparting a feeling of obligation toward these regular marvels.

Integrating natural instruction into school educational programs develops an early attention to the biological significance of fountains and the requirement for their safeguarding. Field trips, intuitive studios, and involved exercises empower understudies to foster a unique interaction with nature and value the excellence and meaning of outpouring environments. Besides, organizations between instructive establishments and protection associations give understudies open doors for involved exploration and preservation projects, sustaining another age of ecological stewards.

Local area outreach drives stretch out training to neighborhood occupants, bringing issues to light about reasonable practices, the effects of human exercises on overflow environments, and the job of networks in protection endeavors. Interpretative signage, directed visits, and intelligent shows at guest focuses further add to public comprehension and enthusiasm for overflow conditions. By cultivating a feeling of ecological proficiency, instruction turns into an impetus for individual and aggregate activities that add to the future progression of fountains.

Strategy Support for Lawful Shields:

Lawful shields are crucial for the future progression of fountains, giving a structure that characterizes passable exercises, safeguards basic natural surroundings, and lays out ramifications for ecological damage. Strategy promotion turns into a fundamental device in forming hearty legitimate systems that focus on the protection and maintainable administration of fountain environments.

Protection associations, in a joint effort with legitimate specialists and natural policymakers, assume a urgent part in supporting for the order and requirement of regulations that defend overflows. This includes taking part in regulative cycles, bringing issues to light about the biological significance of fountains among policymakers, and building public help for protection drives. Compelling arrangement support guarantees that legitimate shields are lined up with contemporary natural difficulties, offering exhaustive insurance for overflow biological systems.

Key components of lawful protections incorporate drafting guidelines that portray regions for preservation, cradle zones to limit human effects, and severe punishments for exercises that undermine the natural respectability of fountains. Natural effect appraisals (EIAs) become required apparatuses for assessing the likely results of advancement projects, guaranteeing that they stick to manageable practices and don't think twice about soundness of fountain biological systems.

Protection Money for Practical Subsidizing:

The reasonable fate of fountains depends on reliable subsidizing for protection drives, research projects, and local area commitment endeavors. Protection finance systems assume a crucial part in guaranteeing the accessibility of assets fundamental for the continuous security and conservation of outpouring environments.

Public and confidential area commitments, awards from altruistic associations, and organizations with corporate elements add to the monetary supportability of protection projects. Laying out devoted assets for overflow preservation, oversaw by legislative offices or non-benefit associations, makes a dependable wellspring of financing for long haul drives. Protection finance systems might incorporate eco-the travel industry income sharing models, biodiversity offset programs, and natural effect securities that influence private speculation for preservation objectives.

Furthermore, worldwide financing instruments, like awards from worldwide ecological assets or associations with global improvement offices, offer essential help for overflows that cross public limits. Protection finance systems should focus on straightforwardness, responsibility, and evenhanded conveyance of assets to guarantee that assets are coordinated toward drives that fundamentally affect safeguarding the magnificence and natural strength of fountains.

Local area Based Checking and Resident Science:

Connecting with nearby networks in observing and research endeavors through local area based checking and resident science drives is a groundbreaking methodology for overflow preservation. Tackling the aggregate information on networks and including them as dynamic members in information assortment contributes significant data as well as encourages a feeling of pride and association with overflow environments.

Local area based checking programs engage nearby occupants to record changes in overflow conditions, report criminal operations, and add to continuous exploration drives. Preparing programs furnish local area individuals with the essential abilities to gather information on water quality, biodiversity, and environment wellbeing. The information gathered through these drives act as significant contributions for logical exploration and illuminate proof based preservation techniques.

Resident science, worked with by computerized stages and versatile applications, broadens the compass of observing endeavors past nearby networks. Anybody with a cell phone or web access can partake in information assortment, adding to a more extensive comprehension of outpouring environments. Resident science drives construct a worldwide local area of people put resources into the preservation of normal marvels, making an organization of backers who champion the future progression of fountains.

The future progression of fountains is a multi-layered story that unfurls at the convergence of natural strength, feasible practices, social safeguarding, versatile procedures, mechanical developments, global joint effort, training, strategy promotion, protection money, and local area based checking. As we explore the intricate territory of natural difficulties and evolving scenes, the basic to get the eventual fate of fountains requires an aggregate responsibility from people, networks, legislatures, and associations around the world.

It is a vow to save the charming magnificence, environmental wealth, and social meaning of fountains for a long time into the future, guaranteeing that these normal miracles keep on enrapturing the hearts and psyches of the people who wonder about their quality.

7.1 Reflect on the future of India's waterfalls in the face of climate change and human impact.

The fate of India's cascades remains at a junction, exploring the unpredictable interchange of environmental change and human effect. These glorious normal marvels, characteristic for the country's different scenes, are confronting remarkable difficulties that compromise their reality and the biological systems they maintain. As we peer into the distance, it is basic to consider the direction that India's cascades could go to and ponder the lengths expected to guarantee their flexibility despite developing ecological elements.

Environmental Change and Hydrological Movements:

Environmental change, portrayed by climbing temperatures, modified precipitation designs, and an expansion in outrageous climate occasions, represents a significant danger to the hydrological cycles that support India's cascades. These fountains, subject to steady water stream, are defenseless to disturbances in precipitation designs, changes in snowmelt elements, and changes in the general accessibility of water. The ramifications stretch out past the tasteful allure of these regular miracles, venturing into the environmental complexities that characterize their biological systems.

The Himalayan cascades, took care of by chilly meltwaters, face the gamble of diminished water accessibility as icy masses retreat. Changes in precipitation designs influence Western Ghats cascades, adjusting the occasional rhythms and possibly prompting variances in stream volumes. Moreover, the Eastern Ghats and the Deccan Level, with their unmistakable cascade environments, are not safe to the effects of environmental change, including adjusted precipitation and expanded temperatures.

Moderating the effect of environmental change on India's cascades requires a thorough methodology that coordinates environment strength into protection techniques. This includes checking hydrological designs, directing weakness appraisals, and carrying out versatile measures to guarantee the supported progression of water. Protection drives should likewise consider the more extensive setting of environmental change, perceiving the interconnectedness of biological systems and the requirement for cooperative endeavors to address this worldwide test.

Human Effect and Anthropogenic Tensions:

The human impression on India's scenes has increased throughout the long term, prompting a scope of anthropogenic tensions that straightforwardly influence the wellbeing and supportability of cascade environments.

Deforestation, urbanization, farming extension, and framework improvement apply aggregate weights on these conditions, changing territories, upsetting natural cycles, and adding to environment fracture.

Deforestation, driven by logging, agribusiness, and metropolitan development, straightforwardly influences the watersheds that feed cascade environments. Loss of woodland cover decreases the limit of scenes to manage water stream, expands the gamble of soil disintegration, and compromises water quality. Urbanization, with its related contamination and change of regular seepage designs, brings pollutants into cascade environments, representing a danger to sea-going life and biological system wellbeing.

Rural works on, including the utilization of pesticides and composts, add to water contamination, influencing both water quality and the biodiversity of outpouring environments. The change of regular scenes into rural fields further lessens the accessible living space for species subject to these conditions.

Foundation advancement, like dams, streets, and the travel industry related offices, can significantly affect cascade biological systems. Dams change regular stream systems, upsetting silt transport and supplement cycling downstream. Streets add to living space discontinuity, hindering the development of species and modifying the normal availability of biological systems. Unregulated the travel industry, on the off chance that not oversaw reasonably, can prompt territory debasement, littering, and unsettling influences that influence both greenery related with cascades.

Tending to human effect on India's cascades requires a change in perspective toward feasible improvement rehearses. Preservation drives should coordinate land-use arranging, advance maintainable horticulture, direct foundation improvement, and execute measures to relieve the unfortunate results of human exercises. Adjusting the requirements of developing populaces with the basic to safeguard the biological respectability of cascade environments requires a deliberate exertion from policymakers, networks, and preservation associations.

Preservation Methodologies for Supportable Fates:

The eventual fate of India's cascades relies on the adequacy of preservation methodologies that address both environmental change effects and human-prompted pressures. These methodologies should be versatile, comprehensive, and grounded in an all encompassing comprehension of the complicated cooperations inside overflow environments.

Environment Strong Protection Arranging:

Preservation arranging should coordinate environment versatility as a center guideline. This includes recognizing key natural surroundings, evaluating weakness to environmental change, and creating procedures to improve biological system flexibility.

Safeguarded regions and support zones assume an essential part in shielding cascade environments, filling in as shelters for species and keeping up with basic biological cycles. The foundation of environment strong hallways works with the development of species, permitting them to adjust to changing ecological circumstances.

Reclamation and Reforestation:

Perceiving the essential job of backwoods in keeping up with watershed wellbeing, reclamation and reforestation endeavors become fundamental. Reforestation programs mean to reestablish regular vegetation in regions that have been deforested or corrupted. Local tree species, adjusted to nearby circumstances, add to soil adjustment, water maintenance, and the making of reasonable natural surroundings for different vegetation. Rebuilding drives ought to likewise think about the renewed introduction of cornerstone species that assume basic parts in biological system elements.

Economical Land Use Practices:

Advancing maintainable land use rehearses is fundamental for moderating human effects on cascade environments. This includes upholding for mindful logging works on, advancing agroforestry drives that incorporate farming and woodland preservation, and carrying out land-use drafting guidelines that focus on the assurance of basic territories. Feasible land use arranging tries to track down a harmony between the necessities of human populaces and the objectives of natural protection.

Local area Commitment and Strengthening:

Networks dwelling close to cascades are vital partners in protection endeavors. Connecting with nearby networks as dynamic members in preservation drives encourages a feeling of pride and obligation. Local area based preservation programs, cooperative dynamic cycles, and the mix of native information add to the maintainability of protection endeavors. Engaging nearby networks with the devices and information to deal with their regular assets makes a common obligation to the conservation of cascade environments.

Guideline of Foundation Advancement:

Managing foundation improvement close to cascades is essential for limiting environmental effects. Ecological effect evaluations (EIAs) ought to be compulsory for projects with the possibility to influence overflow biological systems. Rules for supportable framework advancement, for example, eco-accommodating development rehearses and the foundation of cushion zones, add to limiting the biological impression of human exercises. Compelling guideline requires close joint effort between administrative offices, preservation associations, and the confidential area.

Supportable The travel industry The executives:

The travel industry, whenever oversaw reasonably, can add to nearby economies without compromising the honesty of cascade environments.

Feasible the travel industry rehearses include setting conveying capacities with respect to guest numbers, carrying out trail frameworks to limit living space unsettling influence, and giving natural training to sightseers. Local area based the travel industry drives guarantee that financial advantages are imparted to nearby occupants, cultivating a feeling of obligation for the conservation of regular assets.

Water Asset The board:

Perceiving the significance of water accessibility for cascade environments, water asset the executives turns into a basic part of protection. Reasonable water the executives rehearses incorporate the guideline of water extraction, anticipation of contamination, and support of regular stream systems. Coordinated effort between water asset specialists, protection associations, and neighborhood networks is fundamental for adjusting the requirements of different partners while guaranteeing the natural soundness of outpouring biological systems.

Incorporated Exploration and Observing:

Continuous examination and checking programs give fundamental experiences into the elements of cascade environments. Logical examinations on hydrology, biodiversity, and environment wellbeing illuminate versatile administration procedures. Incorporating conventional natural information with logical examination upgrades the complete comprehension of fountain environments. Long haul checking programs track changes in water quality, species populaces, and living space conditions, empowering preservationists to answer proactively to arising difficulties.

Worldwide Coordinated effort for Shared Water Assets:

A considerable lot of India's cascades are shared across global boundaries, requiring cooperative endeavors for their protection. Two-sided and multilateral arrangements between adjoining nations are fundamental for tending to transboundary challenges. Joint exploration drives, data trade instruments, and facilitated protection procedures add to the common obligation of safeguarding these interconnected environments.

Instruction and Mindfulness Missions:

Training arises as a powerful device for prompting positive change in cultural mentalities toward cascade preservation. Instructive projects in schools, universities, and networks bring issues to light about the natural significance of cascades, the effects of human exercises, and the job of people in preservation endeavors. Interpretative signage, directed visits, and mindfulness crusades at vacationer locations further add to building a preservation disapproved of ethos.

Regulation and Strategy Promotion:

Reinforcing authoritative structures and upholding for strategy changes are basic parts of successful protection. Protection associations and natural backers assume a crucial part in impacting strategy choices that focus on biological safeguarding.

Lawful shields, including rigid punishments for natural infringement, drafting guidelines, and the consolidation of preservation standards into public and local strategies, give an establishment to economical turn of events.

Integrating Native Information:

The mix of native information into protection drives improves the profundity and social significance of conservation endeavors. Native people group frequently have significant hits of knowledge into neighborhood biological systems, occasional examples, and customary land the board rehearses. Cooperative associations that regard and integrate native information add to all encompassing and socially delicate preservation draws near.

7.2 Discuss potential sustainable practices and responsible tourism.

As we face the basic of protecting regular marvels, for example, cascades, coordinating feasible practices and dependable the travel industry becomes urgent. These methodologies act as watchmen against the debasement of environments, guaranteeing that the spectacular excellence of these scenes perseveres for people in the future. In exploring the sensitive harmony between human collaboration and biological conservation, a conversation on expected feasible practices and dependable the travel industry unfurls.

Manageable Practices:

Eco-accommodating Framework Advancement:

Manageable practices start with the preparation and development of foundation encompassing cascades. Eco-accommodating plans that blend with the normal scene limit the natural impression of advancements. Raised walkways, perception stages, and guest focuses made from privately obtained materials embody structures that exist together with the climate instead of forcing upon it.

Reclamation and Reforestation Drives:

Recognizing the critical job of backwoods in supporting cascade biological systems, rebuilding and reforestation drives arise as key parts in the preservation tool compartment. Endeavors to reestablish local vegetation add to soil solidness, water maintenance, and the production of territories fundamental for different greenery. Reforestation programs, directed by standards of biological reclamation, attempt to restore regions influenced by deforestation or debasement.

Economical Land Use and Farming:

Empowering maintainable land use rehearses, especially in the watersheds that feed cascade biological systems, is pivotal. Advancing agroforestry, which coordinates farming with woods protection, keeps up with biological system wellbeing.

Supportable logging rehearses, if fundamental, can be executed, it are not irreversibly adjusted to guarantee that woods environments. Drafting guidelines that focus on the insurance of basic natural surroundings add to keeping up with the environmental respectability of these scenes.

Local area Based Protection Projects:

Engaging neighborhood networks as stewards of cascade environments encourages a feeling of pride and obligation. Local area based protection programs, where occupants effectively take part in dynamic cycles, add to the manageability of safeguarding endeavors. Native information, well established in the social texture of networks, turns into an important resource in forming protection procedures that regard both nature and custom.

Guideline of Water Extraction and Contamination:

Perceiving the significance of water quality for cascade biological systems, guidelines on water extraction and contamination become basic. Supportable water the board rehearses guarantee the upkeep of normal stream systems, forestalling interruptions to oceanic natural surroundings. Forestalling contamination, whether from farming overflow or metropolitan exercises, defends the wellbeing of oceanic environments and the biodiversity they support.

Environment Tough Protection Arranging:

Environment tough protection arranging coordinates versatile procedures that record for the effects of environmental change on cascade biological systems. This includes distinguishing weak regions, executing measures to improve biological system versatility, and working with the development of species through the foundation of environment strong passages. Protection drives that consider the more extensive setting of environmental change add to the drawn out manageability of these scenes.

Mindful The travel industry:

Conveying Limit and Guest The board:

Mindful the travel industry starts with perceiving and sticking to the conveying limit of cascade destinations. Drawing certain lines on guest numbers guarantees that the fragile equilibrium of environments isn't overpowered. Executing guest the executives techniques, for example, coordinated section licenses or reservations, forestalls stuffing and considers a more controlled and economical the travel industry experience.

Instructive Projects and Interpretative Signage:

Instruction shapes the foundation of mindful the travel industry. Carrying out instructive projects in guest habitats, schools, and networks brings issues to light about the biological significance of cascades, the delicacy of these environments, and the job of guests in preservation. Interpretative signage at key areas gives data about the greenery, fauna, and social meaning of the region, cultivating a more profound appreciation for the normal marvels.

Directed Visits and Eco-accommodating Exercises:

Directed visits drove by proficient translators offer guests bits of knowledge into the nature and social history of cascade locales. These visits upgrade the general insight as well as guarantee that guests stick to assigned trails, limiting environment aggravation. Empowering eco-accommodating exercises, for example, bird watching, nature strolls, or photography studios, encourages a more profound association with nature without hurting the climate.

Leave-No-Follow Standards:

Embracing the "leave no follow" standards is central to dependable the travel industry. Guests are urged to do all waste, discard litter dependably, and try not to upset natural life. These standards underscore the significance of limiting the human effect on the climate, permitting ensuing ages to see the value in the scenes in their perfect state.

Local area Commitment and Comprehensive The travel industry:

Mindful the travel industry effectively includes neighborhood networks in the guest experience. Drawing in with nearby occupants, supporting neighborhood organizations, and taking part in local area drove drives add to the feasible improvement of the area. Comprehensive the travel industry rehearses guarantee that monetary advantages are imparted to the networks, encouraging a feeling of shared liability regarding the protection of regular assets.

Volunteer and Preservation Projects:

Offering open doors for travelers to partake in volunteer and protection programs is an effective method for drawing in guests in conservation endeavors. These projects might include natural surroundings rebuilding, untamed life checking, or local area based drives. Vacationers become dynamic supporters of the prosperity of cascade environments, leaving a positive effect on the scenes they have investigated.

Social Awareness and Regard:

Social awareness and regard for neighborhood customs are vital parts of capable the travel industry. Guests are urged to find out about and regard the social legacy of the networks related with cascade destinations. Following social standards, looking for consent for photography, and shunning exercises that might be considered rude add to positive cooperations among guests and neighborhood occupants.

Supporting Preservation Associations:

Mindful vacationers can effectively uphold neighborhood and public protection associations attempting to safeguard cascade environments. Gifts, cooperation in raising money occasions, and chipping in with these associations add to their endeavors in preservation, examination, and local area commitment. Lining up with laid out preservation drives guarantees that travel industry turns into a power for positive change.

Advancing Supportable Transportation:

Empowering the utilization of supportable transportation choices, like public travel or eco-accommodating vehicles, limits the natural effect of movement to cascade locales. Diminishing fossil fuel byproducts related with the travel industry exercises adds to the general supportability of the guest experience.

Criticism and Ceaseless Improvement:

Laying out channels for guest criticism and constant improvement is essential in refining dependable the travel industry rehearses. Normal evaluations of the effect of the travel industry on cascade biological systems, combined with input from guests and nearby networks, take into account versatile administration methodologies. This iterative methodology guarantees that travel industry rehearses advance together as one with changing natural and cultural elements.

Difficulties and Open doors:

While the execution of manageable practices and mindful the travel industry presents a pathway toward the protection of cascade biological systems, challenges endure. Adjusting the financial advantages of the travel industry with the basic of biological conservation requires cautious route. Framework improvement, if unrestrained, can prompt living space debasement. Stuffing and lacking waste administration present dangers to biological system wellbeing. The monetary reliance of nearby networks on the travel industry income requires a fragile balance among protection and financial contemplations.

Be that as it may, inside these difficulties lie potential open doors for advancement and joint effort. Maintainable the travel industry practices can act as impetuses for nearby monetary turn of events, biodiversity protection, and social conservation. Associations between government substances, protection associations, nearby networks, and the confidential area can yield exhaustive techniques that address both the preservation and financial aspects.

7.3 Conclude with a hopeful message about the enduring grace of India's cascades and their importance for generations to come.

In examining the bunch aspects of India's fountains, one can't resist the urge to wonder about the getting through effortlessness that winds through these regular marvels. As the sun sets over the Western Ghats, projecting a brilliant gleam on the lavish vegetation that supports cascades like Run Falls, Athirapally Falls, and Dudhsagar Falls, a significant feeling of wonderment and love saturates the air. These fountains, with their cadenced plummet and ageless presence, murmur accounts of versatility, transformation, and the interconnectedness of all life.

The geographical arrangements that lead to India's shocking cascades are not simple developments of rocks; they are the narrators of the World's excursion.

Cut by the patient hands of time, these fountains stand as demonstrations of the topographical ensemble that has unfurled over centuries. The profound chasms, rough bluffs, and twirling pools tell stories of the World's unique powers, molding scenes that move marvel and thoughtfulness.

Run Falls, in Karnataka, remains as a superb epitome of nature's loftiness. As the Sharavathi Waterway dives in, an entrancing scene unfurls, enamoring all who give testimony. The crude power and magnificence of Run Falls are a sign of the basic powers that have shaped the World's surface, forming the actual forms of our reality.

Wandering into the ethereal domains of Athirapally Falls in Kerala uncovers an outpouring encompassed by rich tropical woodlands. Here, water slides with idyllic effortlessness, making an ensemble of sounds that resound with the energetic biodiversity that calls this district home. Athirapally Falls remains as a demonstration of the sensitive harmony between water, greenery, and fauna — a complex dance that characterizes the biological system's concordance.

Dudhsagar Falls, settled in the Western Ghats of Goa, exemplifies the soul of overflow. The "Ocean of Milk," as its name interprets, plunges from an extraordinary level, its foamy plummet looking like an outpouring of fluid ivory. As the falls combine into an immaculate pool beneath, they mirror the virtue and imperativeness that water bestows to the scenes it graces. Dudhsagar, encompassed by verdant backwoods, represents the soul that supports the rich biodiversity of the Western Ghats.

However, the appeal of India's cascades reaches out past their geographical wonders. It is interlaced with the social embroidery that winds through the country's set of experiences. Nearby legends and stories reinvigorate these fountains, giving them a role as heroes in the stories of networks that have coincided with nature for a really long time. These stories add a layer of persona to the cascades, changing them into social milestones that reverberation the aggregate memory of the land.

Run Falls, for example, holds social importance in Karnataka, with local people crediting heavenly starting points to the falls. Legends discuss the stream's drop as a divine dance, an epitome of Ruler Shiva's vast presentation. Athirapally Falls, then again, is praised in Kerala's old stories as the gathering point of waterways and a consecrated forest — a demonstration of the entwined connection among nature and otherworldliness.

As we dig into the accounts related with Dudhsagar Falls in Goa, we experience stories that resound with the advantageous interaction between human networks and their normal environmental factors. Nearby legends frequently interlaces with the dynamic celebrations celebrated by native clans, making a rich embroidery that highlights the profound association between social legacy and the flowing waters.

While these cascades stand as social milestones, they likewise entice us to examine the unpredictable dance of biological systems that unfurl in their nearness. The Western Ghats, a focal point for cascades in India, protects a biodiversity gold mine. Various plant species, endemic creatures, and a horde of avian life find shelter in the unblemished woodlands encompassing these fountains. The cascades, thus, assume a crucial part in forming microclimates, impacting nearby weather conditions, and giving natural surroundings that encourage biodiversity.

The rich vegetation, biodiversity, and the cooperative connection between the cascades and the biological system structure an enamoring story of concurrence. The plants that stick to the stones, the animals that track down safe-haven in the cool waters, and the birds that wind through the fog — all add to the biological expressive dance that unfurls in the hug of India's fountains.

Notwithstanding, this environmental expressive dance faces difficulties. Human exercises, driven by the requests of advancement and populace development, cast shadows on the holiness of these regular marvels. Deforestation, environment fracture, contamination, and unregulated the travel industry arise as dangers that loom over the unblemished scenes.

The Western Ghats, specifically, gives testimony regarding the fragile exchange among protection and human intercession. As populaces develop and advancement infringes upon these delicate environments, the requirement for reasonable practices and dependable the travel industry becomes central. Adjusting the goals of networks with the basic to protect the biological honesty of cascade environments requires a nuanced and cooperative methodology.

In the Northern locales, where the Himalayan fountains stand tall, an alternate arrangement of difficulties unfurls. The softening of glacial masses, a result of environmental change, represents a danger to the supported progression of water. As these fountains feed into waterways that are helps for millions, the repercussions of frosty retreat reach out a long ways past the prompt scenes.

However, in the midst of these difficulties lie open doors for restoration and change. The eventual fate of India's cascades isn't bound to be an unfortunate account of misfortune, yet rather an account of versatility, transformation, and cooperative stewardship. A story requires an aggregate obligation to offset progress with safeguarding, to embrace supportable practices that orchestrate with the rhythms of nature.

As we ponder the difficulties looked by India's cascades, we find trust in the heap drives and attempts devoted to their preservation. Traditionalists, nearby networks, legislative offices, and travelers the same are arousing to the requirement for a common obligation — an aggregate promise to safeguard and save these normal marvels.

The excursion into the eventual fate of India's cascades welcomes us to imagine a scene where supportable practices are woven into the texture of regular day to day existence. Reforestation endeavors, directed by a pledge to environmental reclamation, revive the woodlands that support these fountains. Feasible land use rehearses, aware of native information, guarantee that human networks flourish without compromising the soundness of biological systems.

In the domain of mindful the travel industry, a change in outlook is in progress. Guests, equipped with mindfulness and a profound appreciation for the sacredness of nature, participate in encounters that leave negligible effect. Instruction programs, interpretative signage, and local area drove drives merge to make a culture of cognizant investigation — one that perceives the fragile harmony between the craving to observe the loftiness of cascades and the basic to safeguard their natural excellence.

A confident story imagines a future where innovation turns into a partner in protection, where headways in checking, research, and maintainable improvement join to make a collaboration among progress and conservation. The incorporation of customary natural information with state of the art research shapes the bedrock of a methodology that regards the insight of ages past while embracing the developments representing things to come.

In the Eastern locales, where lesser-investigated cascades reveal their mysteries, and in the Southern scenes of the Deccan Level, the story proceeds. Here, the fountains cut their ways through different territories, every cascade recounting an interesting story of flexibility and interconnectedness.

As we look into the future, it becomes obvious that the persevering through beauty of India's cascades is definitely not a temporary display yet a developing orchestra. A song reverberates across reality, coaxing us to become caretakers of the charming scenes that have motivated writers, specialists, and visionaries for quite a long time.

For a long time into the future, these fountains will keep on motivating wonderment, reflection, and a profound association with the regular world. Their significance rises above the bounds of topography and culture, turning into a common legacy that joins us chasing a manageable and amicable conjunction with the planet we call home.

In closing this investigation of India's fountains, we are helped that the story to remember cascades is an account of us — of humankind entwined with the multifaceted trap of life. A story welcomes us to be the two observers and watchmen, to wonder about the immortal excellence of cascades while bearing the obligation to save their quintessence.

The getting through effortlessness of India's cascades is a demonstration of the flexibility of nature, the insight of native networks, and the potential for positive change.

As we leave on the excursion into the future, let us convey forward the illustrations murmured by these fountains — the examples of equilibrium, concurrence, and the significant interconnectedness that characterizes the woven artwork of life on The planet.

May the fountains of India keep on flowing down rough scenes and beautiful territories as well as into the hearts and awareness of all who experience their persevering through elegance. In their cadenced plunge, may we track down motivation to fashion an economical and amicable future — a future where the fountains of India stand as encouraging signs for ages on the way.

9 788196 879679